Worksheets

For Classroom or Lab Practice

Edie Carter   Gale Brewer

Amarillo College

5TH EDITION

INTRODUCTORY ALGEBRA

for College Students

BLITZER

PEARSON
Prentice Hall

Upper Saddle River, NJ 07458

Editorial Director, Mathematics: Christine Hoag
Editor-in-Chief: Paul Murphy
Editorial Project Manager: Dawn Nuttall
Assistant Editor: Christine Whitlock
Senior Managing Editor: Linda Mihatov Behrens
Project Manager, Production: Robert Merenoff
Art Director: Heather Scott
Supplement Cover Manager: Paul Gourhan
Supplement Cover Designer: Victoria Colotta
Operations Specialist: Ilene Kahn
Senior Operations Supervisor: Diane Peirano

© 2009 Pearson Education, Inc.
Pearson Prentice Hall
Pearson Education, Inc.
Upper Saddle River, NJ 07458

The author and publisher of this book have used their best efforts in preparing this book. These efforts include the development, research, and testing of the theories and programs to determine their effectiveness. The author and publisher make no warranty of any kind, expressed or implied, with regard to these programs or the documentation contained in this book. The author and publisher shall not be liable in any event for incidental or consequential damages in connection with, or arising out of, the furnishing, performance, or use of these programs.

Printed in the United States of America

10 9 8 7 6 5 4 3 2 1

ISBN-13: 978-0-13-603164-2   Standalone

ISBN-10:   0-13-603164-1   Standalone

ISBN-13: 978-0-13-603165-9   Value Pack

ISBN-10:   0-13-603165-X   Value Pack

Pearson Education Ltd., London
Pearson Education Singapore, Pte. Ltd.
Pearson Education, Canada, Ltd.
Pearson Education—Japan
Pearson Education Australia Pty, Ltd.
Pearson Education North Asia, Ltd.
Pearson Educación de Mexico, S.A. de C.V.
Pearson Education Malaysia, Pte. Ltd.
Pearson Education Upper Saddle River, New Jersey

TABLE OF CONTENTS

Name _____   Date _____

Practice Set 1.1
Introduction to Algebra: Variables and Mathematical Models

Evaluate each algebraic expression for the given value of the variable or variables.

1. $4x - x + 3$ if $x = 2$.   1. _____

2. $2y + y - 4$ if $y = 3$.   2. _____

3. $5x - 4y + 2$ if $x = 4$ and $y = 5$.   3. _____

4. $\dfrac{4x - y}{y - x}$ if $x = 4$ and $y = 6$.   4. _____

Translate each English phrase or sentence into an algebraic expression or equation. Let the variable x represent the number.

5. A number decreased by five.   5. _____

6. The quotient of sixteen and a number.   6. _____

7. Eight less than a number.   7. _____

8. The product of a number and seven increased by four.   8. _____

9. Four less than twice a number is six.   9. _____

10. The sum of a number and twenty is nine.   10. _____

11. Three more than six times a number.   11. _____

12. The difference of a number and ten is equal to twice
     the number.

12. _____

13. Nine added to the quotient of eight and a number.

13. _____

14. Six less a number is twelve.

14. _____

Determine whether the given number is a solution of the equation.

15. $x - 12 = 13; 25$

15. _____

16. $3x + 2 = 14; 4$

16. _____

17. $12y = 60; 5$

17. _____

18. $3a - 6 = 18; 7$

18. _____

19. $4x + 3 = 2x + 11; 4$

19. _____

20. $2(3a - 6) = 4(a - 3); 12$

20. _____

Name _____   Date _____

Practice Set 1.2
Fractions in Algebra

Convert each mixed number to an improper fraction.

1. $4\dfrac{5}{8}$ 1. _____

2. $7\dfrac{3}{16}$ 2. _____

Convert each improper fraction to a mixed number.

3. $\dfrac{44}{5}$ 3. _____

4. $\dfrac{113}{11}$ 4. _____

Write the prime factorization of each number.

5.   80   5. _____

6.   126   6. _____

Reduce each fraction to lowest terms.

7. $\dfrac{12}{8}$ 7. _____

8. $\dfrac{15}{20}$ 8. _____

9. $\dfrac{36}{108}$ 9. _____

Perform the indicated operations.

10. $\dfrac{4}{5} \cdot \dfrac{3}{7}$

10. _____

11. $\dfrac{5}{6} \cdot \dfrac{18}{35}$

11. _____

12. $\dfrac{9}{10} \div \dfrac{3}{4}$

12. _____

13. $18 \div \dfrac{6}{7}$

13. _____

14. $2\dfrac{2}{7} \cdot 4\dfrac{2}{3}$

14. _____

15. $12\dfrac{2}{5} \div 6\dfrac{2}{3}$

15. _____

16. $\dfrac{5}{12} + \dfrac{3}{12}$

16. _____

17. $\dfrac{9}{20} - \dfrac{3}{20}$

17. _____

18. $\dfrac{5}{18} + \dfrac{7}{15}$

18. _____

19. $4\dfrac{3}{8} + 5\dfrac{1}{6}$

19. _____

20. $12\dfrac{7}{8} - 5\dfrac{2}{5}$

20. _____

Practice Set 1.3
The Real Numbers

Express each rational number as a decimal.

1. $\dfrac{5}{8}$ 1. _____

2. $\dfrac{3}{25}$ 2. _____

3. $\dfrac{9}{20}$ 3. _____

Use the set below to answer exercises 4 and 5.

$$\left\{-18,\ -4.3,\ \frac{-1}{3},\ 0,\ 2.8,\ \frac{17}{19},\ \pi,\ 32,\ \sqrt{14}\ \right\}$$

4.   Name the irrational numbers.   4. _____

5.   Name the integers.   5. _____

Use the set below to answer exercises 6 and 7.

$$\left\{-\sqrt{10},\ 15,\ 13\frac{1}{2},\ \frac{3}{8},\ 0,\ -42,\ \sqrt{3},\ 22.7,\ -5.4,\ 1\right\}$$

6.   Name the rational numbers.   6. _____

7.   Name the whole numbers.   7. _____

Use your knowledge of the real numbers to determine whether each statement below is true of false.

8.   A number can be rational and irrational.   8. _____

9.   All natural numbers are also whole numbers.   9. _____

10.   All integers are positive.   10. _____

11.   Whole numbers are sometimes irrational.   11. _____

12.   The absolute value of any negative number is positive.   12. _____

Use the inequality symbols > or < to compare each pair of numbers.

13.   −11.2 ☐   5.3   13. _____

14. $\dfrac{9}{10}$ ☐ $\dfrac{7}{8}$ 14. _____

15. $\dfrac{-5}{6}$ ☐ $\dfrac{-1}{6}$ 15. _____

16.   −6.1 ☐   −8.7   16. _____

17. $\sqrt{10}$ ☐   10   17. _____

18. $|-8|$ ☐ $|6|$ 18. _____

19.   −15 ☐ $|-15|$ 19. _____

20. $|9.2|$ ☐ $|9.35|$ 20. _____

Practice 1.4
Basic Rules of Algebra

For each algebraic expression, (a) give the number of terms and (b) name the like terms, if any, or state that there are none.

1. $4x + 12$

1a. _____

b. _____

2. $7a - 12 + 3a$

2a. _____

b. _____

3. $5y + 8 + 2y + 9z$

3a. _____

b. _____

4. $a + 5a - 10 + 4a + 3b$

4a. _____

b. _____

State the property being illustrated in each example.

5. $6(5x) = (6 \cdot 5)x$

5. _____

6. $14 + x + 8 = 14 + 8 + x$

6. _____

7. $9(y + 4) = 9y + 36$

7. _____

8. $(a + 100) + 50 = a + (100 + 50)$

8. _____

9. $\dfrac{1}{4}(8x - 20) = 2x - 5$

9. _____

10. $25 \cdot y \cdot 7 = 25 \cdot 7 \cdot y$

10. _____

Simplify each algebraic expression.

11. $9y - y$ 11._____

12. $11x - 8x$ 12._____

13. $8(a + 5)$ 13._____

14. $6(2x + 3) - 5x + 2$ 14._____

15. $\dfrac{1}{3}(9x + 15) + 3x$ 15,_____

16. $2(5y + 4) + 3(8 - 2y)$ 16._____

Write each English phrase as (a) an algebraic expression. then (b) simplify. Let the variable x represent the number.

17.   The difference of eight times a number and twice the number.   17a._____

                                                                                                                 b._____

18.   Five times the product of a number and four.   18a._____

                                                                                                                 b._____

19.   The sum of twice a number and sixteen decreased by five.   19a._____

                                                                                                                 b._____

20.   Eight times the sum of a number and four decreased by three times the number.   20a._____

                                                                                                                 b._____

Practice Set 1.5
Addition of Real Numbers

Find each sum.

1. $18 + (-14)$ 1. _____

2. $-32 + 32$ 2. _____

3. $-51 + 27$ 3. _____

4. $-13 + (-25)$ 4. _____

5. $-5.8 + 3.6$ 5. _____

6. $\dfrac{-3}{10} + \dfrac{2}{10}$ 6. _____

7. $26 + (-84)$ 7. _____

8. $-50 + (-50)$ 8. _____

9. $-\dfrac{1}{8} + -\dfrac{3}{8}$ 9. _____

10. $-0.53 + 0.42$ 10. _____

11. $15 + (-6) + (-4)$

12. $-62 + 41 + 13$

12._____

13. $20 + (-10) + (-13) + 19$

13._____

14. $-5.6 + 3.7 + (-2.7) + 5.1$

14._____

15. $-18 + \left(-\dfrac{3}{7}\right) + 22 + \left(\dfrac{-4}{7}\right)$

15._____

Simplify each algebraic expression.

16. $15x + (-18x)$

16._____

17. $-9a + 3a + 7a$

17._____

18. $-27y + 16y + y$

18._____

19. $-8x + 10 + 8x + (-10)$

19._____

20. $-3(x + 4) + 2(x + 5)$

20._____

Practice Set 1.6
Subtraction of Real Numbers

Subtract.

1. $8 - 14$

1. _____

2. $10 - (-12)$

2. _____

3. $-5.3 - 12$

3. _____

4. $420 - 532$

4. _____

5. $57 - (-13)$

5. _____

6. $\dfrac{-7}{12} - \left(\dfrac{-5}{12}\right)$

6. _____

7. $\dfrac{-4}{9} - \dfrac{4}{9}$

7. _____

8. $-212 - 33$

8. _____

9. $-35 - 15 - 10$

9. _____

10. $14 - 12 - (-16)$

10. _____

Simplify each series of additions and subtractions.

11. $15 - 12 - (-17)$ 11. _____

12. $-8 - 14 + 13 - (-7)$ 12. _____

13. $115 - (-123) - 101 + 98$ 13. _____

14. $\dfrac{-1}{4} - \dfrac{1}{6} - \left(\dfrac{-1}{8}\right)$ 14. _____

15. $-310 - 42 - 55 + 123 - (-72)$ 15. _____

Simplify each algebraic expression.

16. $5y - 8 - y + 12$ 16. _____

17. $12m - 6m - 17 - (-3m) - 5$ 17. _____

18. $4x - (-16) - 6x + 12 + 5x$ 18. _____

19. $4(x + 2) - 3(x + 4)$ 19. _____

20. $7(3 - y) + 12(y - 2)$ 20. _____

Name _____   Date _____

Practice Set 1.7
Multiplication and Division of Real Numbers

Perform the indicated operations, simplify if possible.

1. $(3)(-4)$

1. _____

2. $-24 \div -6$

2. _____

3. $(-2)(5)(-8)(15)(0)(-5)$

3. _____

4. $-14 \div 0$

4. _____

5. $(-3)(4)(-1)(-5)$

5. _____

6. $\dfrac{-12}{5} \div \dfrac{6}{10}$

6. _____

7. $\left(-\dfrac{3}{4}\right)\left(-\dfrac{6}{5}\right)$

7. _____

8. $8y - 9y$

8. _____

9. $-5\left(\dfrac{-2}{5}a\right)$

9. _____

10. $2(x + 5) - 4(2x - 3)$

10. _____

11. $(-0.8)(0.12)$ 11. _____

12. $-5\left(-\dfrac{1}{5}\right)$ 12. _____

13. $-x + x$ 13. _____

14. $7x + 3(x + 4)$ 14. _____

15. $22.2 \div -2$ 15. _____

16. $-50 \div -10$ 16. _____

17. $(4)(-3)(-1)(2)$ 17. _____

18. $-3(x - 6)$ 18. _____

19. $-(5y - 4)$ 19. _____

20. $-3 \div \left(\dfrac{6}{5}\right)$ 20. _____

Practice Set 1.8
Exponents and Order of Operations

Evaluate each exponential expression.

1. -7^2 1. _____

2. $(-4)^3$ 2. _____

3. $(-8)^2$ 3. _____

4. -1^4 4. _____

5. 5^3 5. _____

Simplify each algebraic expression, if possible.

6. $5x^2 - 3 - 2x^2$ 6. _____

7. $9x^2 - 3x + 8$ 7. _____

8. $3y^2 + 2y - y^2 + 4y + 7$ 8. _____

9. $4(x+7) - 3[2(x-4)]$ 9. _____

10. $5(x^2+1) - 2(2-x^2)$ 10. _____

Name _____   Date _____

Simplify each expression using the order of operations.

11. $3^2 + 2^3$

11. _____

12. $(4+2)^2 + (3-5)^2$

12. _____

13. $14 - 2 \cdot 3 - 8$

13. _____

14. $6^2 \div 2 \cdot 3 - 3^3$

14. _____

15. $\dfrac{5 + 3^2}{1 - (-1)}$

15. _____

16. $5[10 - (2^3 - 20)]$

16. _____

17. $24 \div 4 \cdot 2 + 5 \cdot 3$

17. _____

18. $\dfrac{(-2)^3 - 2^3}{4^2 \div 8}$

18. _____

19. $5(-3) - |-16 \div 2 \cdot 4|$

19. _____

20. $\left[\dfrac{-5}{8} - \dfrac{3}{8}\right]\left[\dfrac{-3}{4} + \dfrac{6}{8}\right]$

20. _____

Practice Set 2.1
The Addition Property of Equality

Identify each equation as linear or not linear.

1. $x + 7 = 15$ 1. _____

2. $x^2 - 3 = 9$ 2. _____

3. $\dfrac{11}{x} = 4$ 3. _____

4. $\sqrt{5}x + \pi = 0.\overline{3}$ 4. _____

5. $|x + 1| = 7$ 5. _____

Solve each equation using the addition property of equality. Be sure to check your proposed solutions.

6. $x - 9 = -11$ 6. _____

7. $-6 = x + 9$ 7. _____

8. $-8 + x = -8$ 8. _____

9. $18 = x + 12$ 9. _____

10. $x + 20 = 10$ 10. _____

11. $4.5 + x = 4.5$ 11. _____

12. $x + 11.4 = 12$ 12. _____

13. $x - 1 = -\dfrac{1}{2}$ 13. _____

14. $3.5 = x + 2$ 14. _____

15. $x + \dfrac{7}{2} = \dfrac{13}{2}$ 15. _____

16. $2x = x + 5$ 16. _____

17. $7x - 6 = 6x - 1$ 17. _____

18. $2(x - 1) = x$ 18. _____

19. $7(x + 4) = 6x + 24$ 19. _____

20. $4(x - 2) = 3(x + 1)$ 20. _____

Practice Set 2.2
The Multiplication Property of Equality

Solve each equation using the multiplication property of equality. Be sure to check your proposed solution.

1. $\dfrac{x}{6} = -2$ 1. _____

2. $-4x = 12$ 2. _____

3. $\dfrac{2}{3}x = 12$ 3. _____

4. $-12x = -60$ 4. _____

5. $-x = -5$ 5. _____

6. $\dfrac{3}{4}x = 6$ 6. _____

7. $-10x = 5$ 7. _____

8. $4x = 0$ 8. _____

9. $25 = \dfrac{-5}{7}x$ 9. _____

10. $-8 = \dfrac{1}{2}x$ 10. _____

Name _____   Date _____

Solve each equation using both the addition and multiplication properties of equality. Be sure to check your proposed solutions.

11. $25 + x = 5x + 1$ 11. _____

12. $4x + 8 = 16$ 12. _____

13. $3x + 7 = 5x - 13$ 13. _____

14. $a - 3 = 3a + 15$ 14. _____

15. $a + 10 = 3a - 4$ 15. _____

16. $-x - 4 = 4$ 16. _____

17. $5x + 1 = 2x - 5$ 17. _____

18. $2z + 7 = 8z + 1$ 18. _____

19. $6x - 4 = 4x + 7$ 19. _____

20. $x + 9 = 3x + 15$ 20. _____

Practice Set 2.3
Solving Linear Equations

Solve each equation. Be sure to check your proposed solution by substituting it for the variable in the original equation.

1. $2(x - 2) = 5(x + 1)$ 1. _____

2. $6x + 2x - x = 10 + 4$ 2. _____

3. $3(x + 4) = 5(x - 2)$ 3. _____

4. $2(6x - 4) = -2$ 4. _____

5. $2x + 4 = x + 4$ 5. _____

6. $3x + 2x - 8x = -27$ 6. _____

7. $3(4x + 5) = -33$ 7. _____

8. $4x - (2x + 5) = 11$ 8. _____

9. $2(x + 4) - 5x = 4(x - 5)$ 9. _____

10. $5(x - 3) = 3(2x - 4)$ 10. _____

11. $7x - (2x + 4) = 3(x - 6)$ 11. _____

12. $-9(x + 3) = 4x - (x + 3)$ 12. _____

13. $2(3x+1) = 2(5x-1)$ 13. _____

14. $-18-3(x-4) = -2x-6$ 14. _____

15. $9(2-y)+5 = 4y-16$ 15. _____

Solve and check each equation. Begin your work by rewriting each equation without fractions.

16. $\dfrac{x}{3} - 2 = 8$ 16. _____

17. $\dfrac{x}{2} - \dfrac{5}{3} = \dfrac{3x}{4} - \dfrac{1}{6}$ 17. _____

18. $\dfrac{x-2}{4} + 1 = \dfrac{x+4}{3}$ 18. _____

Solve each equation. Use words or set notation to identify equations that have no solution or equations that are true for all real numbers.

19. $2(x+4) = 2x-8$ 19. _____

20. $2x+4x-5x+11 = x+13-2$ 20. _____

Practice Set 2.4
Formulas and Percents

Solve each formula for the specified variable.

1. $P = a + b + c$ for a

   1. _____

2. $V = lwh$ for l

   2. _____

3. $y = mx + b$ for x

   3. _____

4. $\dfrac{A}{w} = l$ for w

   4. _____

5. $\dfrac{2A}{b} = h$ for A

   5. _____

6. $A = \dfrac{1}{2}h(a+b)$ for h

   6. _____

7. $PV = nRT$ for R

   7. _____

8. $Ax + By = C$ for B

   8. _____

Express each percent as a decimal.

9. 15%

   9. _____

10. 9.25%

    10. _____

11. $\dfrac{3}{4}\%$

    11. _____

Express each decimal as a percent.

12.   0.71   12. _____

13.   0.015   13. _____

14.   49   14. _____

Use the percent formula, $A = PB$; A is P percent of B to solve exercises 15-20.

15.   What is 7% of 100?   15. _____

16.   8 is 40% of what?   16. _____

17.   16% of what number is 40?   17. _____

18.   18 is what percent of 72?   18. _____

19.   What percent of 7.5 is 1.125?   19. _____

20.   If 45 is decreased to 36, the decrease is what percent   20. _____
       of the original number?

Practice Set 2.5
An Introduction to Problem Solving

Let x represent the number. Use the given conditions to (a) write an equation and then (b) solve the equation to find the number.

1.   A number decreased by five is eleven.   1a. _____

                                                                      b. _____

2.   The quotient of five and a number is negative five.   2a. _____

                                                                      b. _____

3.   Three more than twice a number is five.   3a. _____

                                                                      b. _____

4.   Eight less than a number is eighteen.   4a. _____

                                                                      b. _____

5.   A number increased by two is three times the number.   5a. _____

                                                                      b. _____

6.   Twice the difference of a number and four is six.   6a. _____

                                                                      b. _____

7.   Eleven is the same as a number less four.   7a. _____

                                                                      b. _____

8.   Three times a number increased by four is negative eight.   8a. _____

                                                                      b. _____

9.   Two less than six times a number is sixteen.   9a._____

                                                                     b._____

10. The difference of a number and four is ten.   10a._____

                                                                     b._____

11.   Nine more than the product of eight and a number is one.   11a._____

                                                                     b._____

12. Three times the sum of a number and five is   12a._____
     negative twenty-one.
                                                                     b._____

13. Six more than two times a number is that number   13a._____
     less eleven.
                                                                     b._____

14. If the quotient of twice a number and three is decreased   14a._____
     by four, the result is negative ten.
                                                                     b._____

15. Four times a number less eight is the same as twice a   15a._____
     number increased by four.
                                                                     b._____

Name _____   Date _____

Practice Set 2.6
Solving Linear Inequalities

(a) Graph the solutions of each inequality on a number line and then (b) express the solution set of each inequality in interval notation.

1. $x \geq -4$

    1a.

    b. _____

2. $x < 2$

    2a.

    b. _____

3. $x > 3$

    3a.

    b. _____

4. $x \leq \dfrac{1}{2}$

    4a.

    b. _____

5. $-2 < x \leq 4$

    5a.

    b. _____

6. $0 \leq x \leq 3$

    6a.

    b. _____

Use the addition property of inequality to solve each inequality. (a) Graph the solutions of each inequality on a number line and then (b) express the solution set of each inequality in interval notation.

7. $x + 4 > 7$

    7a.

    b. _____

27

8. $x - 3 \leq 2$

8a. _____

b. _____

9. $4x + 9 > 3x + 4$

9a. _____

b. _____

10. $x + \dfrac{1}{2} < \dfrac{1}{8}$

10a. _____

b. _____

11. $4x + 2 \geq 3 + 3x$

11a. _____

b. _____

Use the multiplication property of inequality to solve each inequality. (a) Graph the solutions of each inequality on a number line and then (b) express the solution set of each inequality in interval notation.

12. $\dfrac{1}{2}x > -2$

12a. _____

b. _____

13. $\dfrac{x}{4} \leq 1$

13a. _____

b. _____

14. $3x \geq 9$

14a. _____

b. _____

15. $-4x < -12$

15a. _____

b. _____

16. $-x \geq 3$

16a. _____

b. _____

Use both the addition and multiplication properties of inequality to solve each inequality.
(a) Graph the solutions of each inequality on a number line and then (b) express the solution set of each inequality in interval notation.

17. $-(x + 6) > -5$

17a.

   b. _____

18. $2(4x + 1) \leq -6$

18a

   b. _____

19. $4x + 9 > 2x + 3$

19a.

   b. _____

20. $6(x - 4) \geq 3(x - 5)$

20a.

   b. _____

Solve each inequality.

21. $6x < 4 + 6x$

21. _____

22. $2x \geq 2x + 6$

22. _____

Name _____   Date _____

Name _____   Date _____

Practice Set 3.1
Further Problem Solving

Solve each application problem.

1. Six thousand dollars is deposited into a savings account that   1. _____
    has a 2.5% annual interest rate. Find the interest earned at the
    end of the first year.

2. If $3,010 is invested in a mutual fund paying 4.2% annual   2. _____
    interest, find the interest earned after 2 years.

3. You invest $10,000 in two different accounts. One account   3. _____
    earns 4% annual interest and the other earns 5% annual
    interest. If the interest earned after one year is $455, find
    how much was invested in each account.

4. Together, Jody and Suzette had $14,000 to invest. Jody chose   4. _____
    to invest his money in a mutual fund paying 6% interest.
    Suzette invested her money in a stock paying 5.5% interest.
    If their money earned a combined total of $798 in interest
    after one year, find out how much each person invested.

5. You invested $2,100 in two different accounts paying 8% and   5. _____
    9.5% annual interest respectively. If the total interest earned
    after one year is $186, how much was invested at each rate?

6. A 16.9 ounce household fabric spray container contains 20%   6. _____
    alcohol. How much alcohol is in the container?

7. A 750 ml bottle of champagne contains 9.5% alcohol. How   7. _____
    much alcohol is in the bottle?

8. The Lions Club is raising money by selling mixed nuts at the   8. _____
    local market on Saturdays. They will mix two different
    brands of mixed nuts together. One brand is 80% peanuts
    and the other is 20% peanuts. How many pounds of each
    brand will need to be mixed together to get 25 pounds of the
    mixed nuts so that the mixture is only 44% peanuts?

31

9. How many gallons of 30% alcohol solution and 70% alcohol   9. _____
    solution must be mixed to get 16 gallons of 50% solution?

10. How much 15% antifreeze and 55% antifreeze should be   10. _____
    mixed to get 50 gallons of 30% antifreeze?

11. Jack is riding his bike at a steady rate of 12 miles per hour.   11. _____
    How long will it take him to go a distance of 66 miles?

12. An airplane can fly a distance of 1,666 miles in 3 1/2 hours.   12. _____
    How fast is the plane flying?

13. Two airplanes leave Dallas, Texas at the same time going in   13. _____
    opposite directions. One is traveling at 880 km/h and the
    other at 820 km/h. How long will it be until the planes are
    10,200 km apart?

14. Two ships leave port at the same time, one traveling north and   14. _____
    one south. One ship is traveling at 45 mph and the other
    12 mph slower. How long will it take for the ships to be
    936 miles apart?

15. Two families live 350 miles apart. They are planning to meet   15. _____
    for lunch on a Saturday afternoon. Each family leaves home
    at the same time. One travels 20 mph less than twice as fast
    as the other, and they will meet in 5 hours. How fast is each
    family traveling?

Name _____   Date _____

Practice Set 3.2
Ratio and Proportion

Express each ratio as a fraction in lowest terms.

1. 32 to 48   1. _____

2. 20 to 55   2. _____

3. 76 to 16   3. _____

Determine whether each of the following is a proportion by using cross products.

4. $\dfrac{8}{9} = \dfrac{20}{22}$ 4. _____

5. $\dfrac{15}{16} = \dfrac{180}{192}$ 5. _____

Solve each proportion.

6. $\dfrac{x}{5} = \dfrac{105}{75}$ 6. _____

7. $\dfrac{4.2}{a} = \dfrac{8}{5}$ 7. _____

8. $\dfrac{18}{30} = \dfrac{m}{4}$ 8. _____

9. $\dfrac{p}{15} = \dfrac{11}{12}$ 9. _____

10. $\dfrac{x}{54} = \dfrac{5}{6}$ 10. _____

11. $\dfrac{x}{4} = \dfrac{x-3}{2}$ 11. _____

12. $\dfrac{y+1}{5} = \dfrac{3y}{10}$ 12, _____

13. $\dfrac{2x-3}{x+5} = \dfrac{5}{9}$ 13. _____

Use a proportion to solve each problem.

14.   An office building leases space monthly for $32.50 for each   14. _____
       50 square foot. How much would it cost to lease a 1,050
       square foot space?

15.   A store advertises 4 oversized bath towels on sale for $25.   15. _____
       How much would it cost to purchase 10 bath towels?

16.   In a newspaper survey 4 out of 5 people that responded said   16. _____
       they preferred bottled water to the local tap water. If 40,000
       people responded, how many preferred bottled water?

17.   The ratio of female nurses to male nurses at Mercy General   17. _____
       Hospital is 7 to 2. If there are 405 nurses on staff, how many
       are males? How many are females?

18.   One instructor's pass rate for her basic algebra classes is 85%.   18. _____
       This means that 17 out of every 20 of her students will pass.
       One semester she has 160 enrolled in her classes. How many
       could be expected to pass her class?

Practice Set 3.3
Problem Solving in Geometry

Use the formulas for perimeter, area, circumference and volume to solve the problems.

1.   Find the area of a rectangle with a length of 16 inches and a   1. _____
      width of 7 inches.

2.   Find the perimeter of a rectangle with a length of 24 centimeters   2. _____
      and a width of 10 centimeters.

3.   Find the area of a triangle that has a base 10 inches and a   3. _____
      height of 5 inches.

4.   Find the length of a rectangle in which the width is 8 meters   4. _____
      and the area is 116 m^2.

5.   Find the area of a circle that has a radius of 3 inches. Express   5. _____
      your answer in terms of π, then round to the nearest whole
      number.

6.   Find the area of a circle with a diameter of 14 millimeters.   6. _____
      Express your answer in terms of π, then round to the nearest
      whole number.

7.   Find the circumference of a circle with a radius of 5   7. _____
      centimeters. Express your answer in terms of π, then round to
      the nearest whole number.

8.   Find the circumference of a circle with a diameter of   8. _____
      8 decimeters. Express your answer in terms of π, then
      round to the nearest whole number.

9.   Find the radius of a circle that has a circumference of 18π feet. 9. _____

10.   Find the diameter of a circle that has an area of $49\pi \text{ yd}^2$. 10. _____

11.   Find the volume of a cube with a length of 2.5 inches.   11. _____

12.   Find the volume of a rectangular solid with a length of 7 inches, a width of 5.5 inches and a height of 3 inches.   12. _____

13.   Find the volume of a circular cylinder with a height of 3 feet and a radius of 1.5 feet. Express your answer in terms of π, then round to the nearest whole number.   13. _____

14.   Find the volume of a ball if the diameter is 12 inches. Express your answer in terms of π then round to the nearest whole number.   14. _____

15.   The largest angle of a triangle is six more than three times the smallest angle. The third angle is twice the smallest. Find the measure of each angle.   15. _____

16.   One angle of a triangle is twice the first angle. The third angle is eight more than the first angle. Find the measure of each angle.   16. _____

17.   Two angles are complementary. One measures 73°. What is the measure of the other angle?   17. _____

18.   Two angles are supplementary. One angle is 42°. Find the measure of its supplement.   18. _____

19.   Two angles are supplementary. One angle measures x and the other $3x + 4$. Find the measure of each angle.   19. _____

20.   Two angles are complementary. One angle measures x and the other $2x + 3$. Find the measure of each angle.   20. _____

Name _____   Date _____

Practice Set 4.1
Graphing Equations in Two Variables

Plot the given point in a rectangular coordinate system. Indicate which quadrant each point lies.

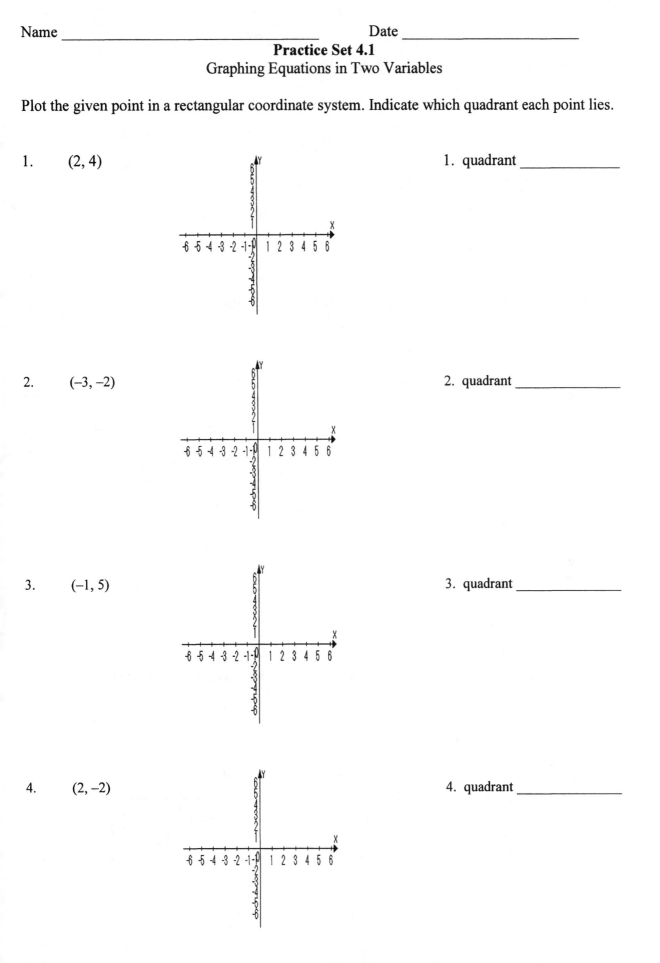

1.   (2, 4)   1. quadrant _____

2.   (−3, −2)   2. quadrant _____

3.   (−1, 5)   3. quadrant _____

4.   (2, −2)   4. quadrant _____

37

Name _____   Date _____

Give the ordered pairs that correspond to the point labeled in the figure.

5. A _____

6. B _____

7. C _____

8. D _____

Determine whether each ordered pair is a solution of the given equation.

9. $y = -2x$

    a. $(-2, -4)$ 9a. _____

    b. $(0, -2)$ b. _____

    c. $(1, -2)$ c. _____

10. $2x + 5y = 10$

    a. $(5, 0)$ 10a. _____

    b. $(0, -2)$ b. _____

    c. $(5, -4)$ c. _____

Complete each table of values for each given equation. Graph the ordered pairs from the table.

11. $y = x$

x	$y = x$	(x, y)
-2		
-1		
0		
1		
2		

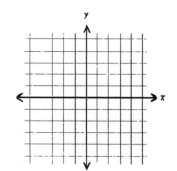

12. $y = -2x - 1$

x	$y = -2x - 1$	(x, y)
−2		
−1		
0		
1		
2		

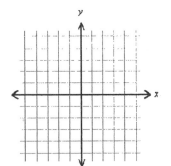

13. $y = 3x - 1$

x	$y = 3x - 1$	(x, y)
−2		
−1		
0		
1		
2		

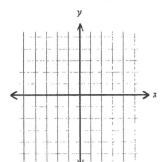

14. $y = 2$

x	$y = 2$	(x, y)
−2		
−1		
0		
1		
2		

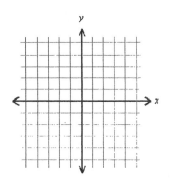

15. $y = -\dfrac{1}{2}x + 1$

x	$y = -\dfrac{1}{2}x + 1$	(x, y)
−4		
−2		
0		
2		
4		

Name _____   Date _____

Name _____   Date _____

Practice Set 4.2
Graphing Linear Equations Using Intercepts

Find the *x*-intercept and the *y*-intercept of the graph of each equation. Do not graph the equation.

1. $5x - 3y = 15$ 1. _____

2. $2x - y = 4$ 2. _____

3. $x + y = 0$ 3. _____

4. $3x + 2y = -6$ 4. _____

5. $4x - 3y = 7$ 5. _____

Use intercepts and a check point to graph each equation.

6. $x + y = 4$ 7. $2x - 4y = 8$

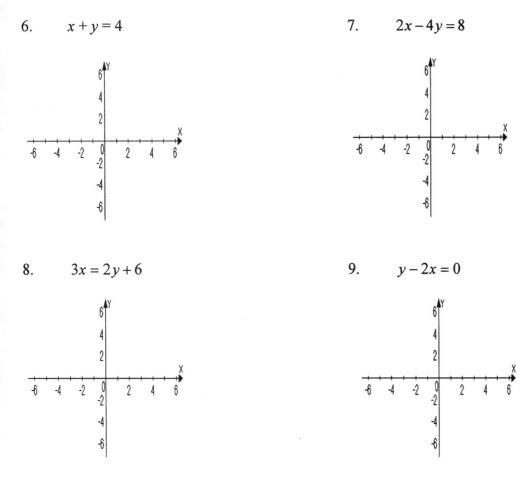

8. $3x = 2y + 6$ 9. $y - 2x = 0$

10. $2x + 3y = 5$

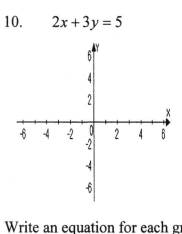

Write an equation for each graph.

11.

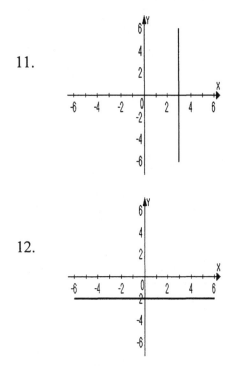

11. _____

12.

12. _____

Graph each equation.

13. $x + 2 = 0$

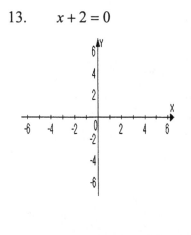

14. $2y - 6 = 0$

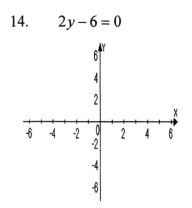

Name _____   Date _____

Practice Set 4.3
Slope

(a) Find the slope of the line passing through each pair of points or state that the slope is undefined. (b) Then indicate whether the line through the point rises, falls, is horizontal, or is vertical.

1.   (1, 4) (2, 7)   1a. _____

                                                       b. _____

2.   (2, 4) (3, 4)   2a. _____

                                                       b. _____

3.   (–3, 5) (2, –8)   3a. _____

                                                       b. _____

4.   (1, 6) (1, 5)   4a. _____

                                                       b. _____

5.   (2, 1)(4, –3)   5a. _____

                                                       b. _____

6.   (0, 1)(–3, –2)   6a. _____

                                                       b. _____

7.   (2, 8)(2, –8)   7a. _____

                                                       b. _____

8.   (4, 1)(–5, 1)   8a. _____

                                                       b. _____

Find the slope of each line or state that the line is undefined.

9.   9. _____

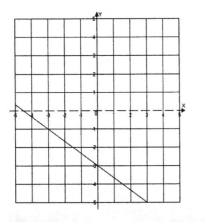

43

10.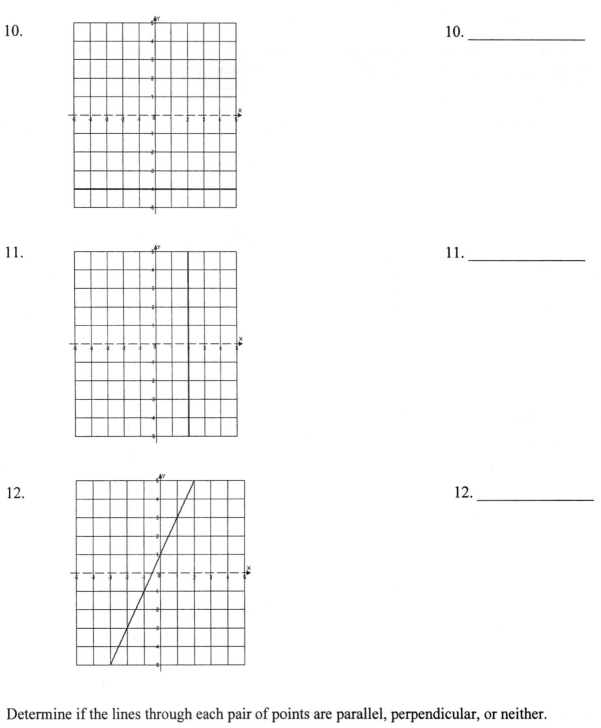

10. _____

11.

11. _____

12.

12. _____

Determine if the lines through each pair of points are parallel, perpendicular, or neither.

13.   (3, –4)(2, 8) and (4, 6)(–8, 5)   13. _____

14.   (8, –1)(4, –3) and (–5, 5)(–3, 6)   14. _____

15.   (2, –1)(4, 5) and (–6, 1)(2, 7)   15. _____

Name _____   Date _____

Practice Set 4.4
The Slope-Intercept Form of the Equation of a Line

Find the (a) slope and (b) the y-intercept of the line with the given equation.

1. $y = 2x - 4$

1a. _____

b. _____

2. $y = -\dfrac{1}{2}x$

2a. _____

b. _____

3. $y = 3$

3a. _____

b. _____

4. $y = 6 - x$

4a. _____

b. _____

Begin by solving the linear equation for y. This will put the equation in slope-intercept form. Then find the (a) slope and (b) the y-intercept of the line with the equation. (c) Graph the equation using the slope and y-intercept.

5. $y = \dfrac{2}{5}x + 3$

5a. _____

b. _____

6. $-2x + y = -1$

6a. _____

b. _____

7.　　$-3x - 4y = 0$　　　　7a. _____

　　　　　　　　　　　　　　　b. _____

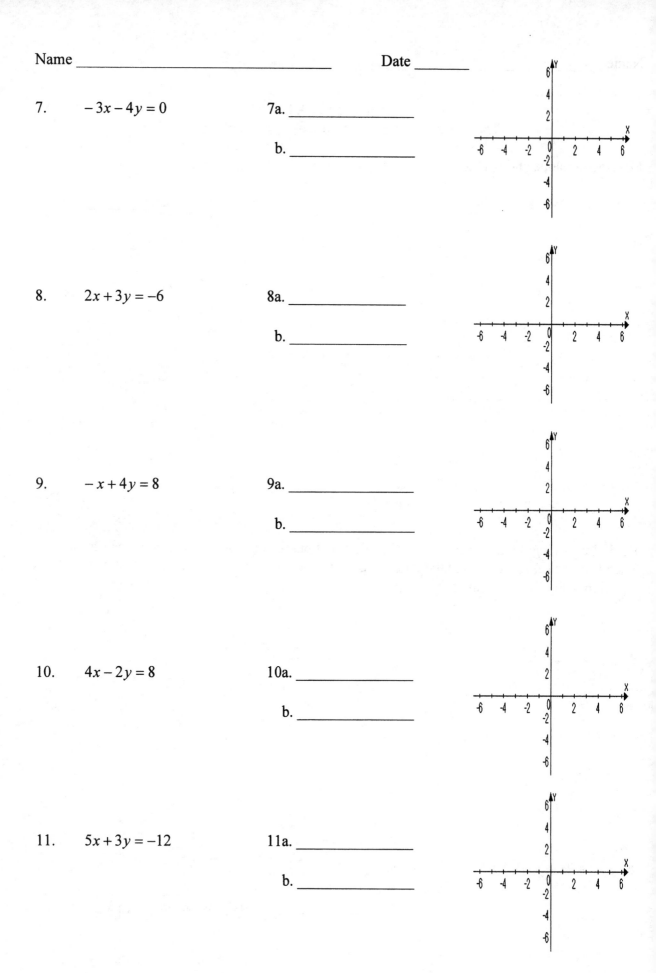

8.　　$2x + 3y = -6$　　　　8a. _____

　　　　　　　　　　　　　　　b. _____

9.　　$-x + 4y = 8$　　　　9a. _____

　　　　　　　　　　　　　　　b. _____

10.　　$4x - 2y = 8$　　　　10a. _____

　　　　　　　　　　　　　　　b. _____

11.　　$5x + 3y = -12$　　　11a. _____

　　　　　　　　　　　　　　　b. _____

Name _____   Date _____

Fill in the blanks.

12. A line with a _____ slope slants up from left to right.

13. A line with a _____ slope slants down from left to right.

14. A horizontal line has _____ slope.

15. A vertical line has _____ slope.

Name _____   Date _____

Practice Set 4.5
The Point-Slope Form of the Equation of a Line

Write (a) the point slope of the equation of the line satisfying each of the conditions. Then use the point slope form of the equation to (b) write the slope-intercept form of the equation.

1.   slope –2 passing through (3,1)

1a. _____

b. _____

2.   passing through (1, 4) (–2, –2)

2a. _____

b. _____

3.   slope $\dfrac{-2}{3}$ passing through (0, 1)

3a. _____

b. _____

4.   passing through (2, –1) (4, –1)

4a. _____

b. _____

5.   slope –4 passing through the origin

5a. _____

b. _____

6.   slope –1 passing through (–3, 2)

6a. _____

b. _____

7.   passing through (0, 4) (3, 7)

7a. _____

b. _____

8.   slope –2 passing through (1, 0)

8a. _____

b. _____

9.   slope $\dfrac{1}{2}$ passing through (–2, 4)

9a. _____

b. _____

10.   x-intercept 3 and y-intercept –1

10a. _____

b. _____

Practice Set 4.6
Linear Inequalities in Two Variables

Determine whether each ordered pair is a solution of the given equality.

1. $x + y \geq -3$

    a. $(1, -4)$ 1a. _____

    b. $(-3, 3)$ b. _____

    c. $(-1, -5)$ c. _____

2. $2x - y < 4$

    a. $(3, 2)$ 2a. _____

    b. $(0, -5)$ b. _____

    c. $(1, 3)$ c. _____

3. $y \leq -x + 1$

    a. $(0, 0)$ 3a. _____

    b. $(1, 4)$ b. _____

    c. $(2, -4)$ c. _____

4. $y > -3x - 2$

    a. $(1, 5)$ 4a. _____

    b. $(0, -1)$ b. _____

    c. $(2, -8)$ c. _____

Graph each inequality.

5. $y < -2$

6. $3x - 2y \geq -4$

7. $x - y < 3$

8. $x > -2$

9. $6x - 3y > -3$

10. $x + y \leq 3$

11. $3x + y \leq 3$

12. $4x + 6y > 6$

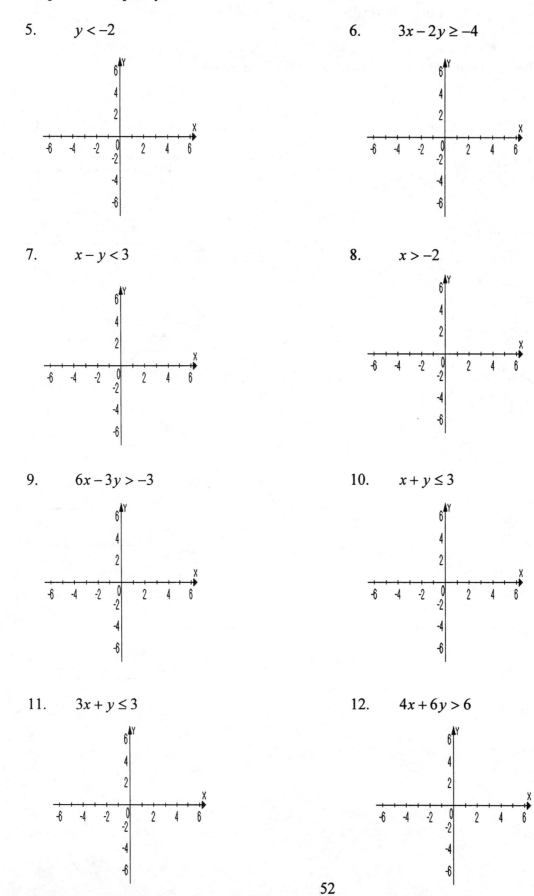

Name _____   Date _____

Practice Set 5.1
Solving Systems of Linear Equations by Graphing

Determine whether the given ordered pair is a solution to the system.

1. $x + y = 9$ (4, 5)   1. _____
    $4x - y = 11$

2. $3x - 2y = 3$ $\left(\dfrac{2}{3}, \dfrac{-1}{2}\right)$ 2. _____

    $6x + 4y = 2$

3. $x - 5y = 11$ (6, –1)   3. _____
    $3x - y = 17$

4. $2x - 7y = -4$ (5, 2)   4. _____
    $y = -2x + 8$

Solve the following systems by graphing. Use set notation to express solution sets. If there is no solution or an infinite number of solutions, so state.

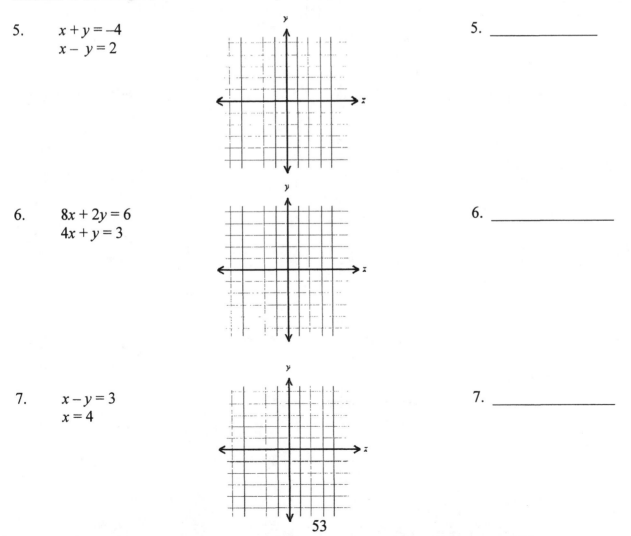

5. $x + y = -4$ 5. _____
    $x - y = 2$

6. $8x + 2y = 6$ 6. _____
    $4x + y = 3$

7. $x - y = 3$ 7. _____
    $x = 4$

Name _____   Date _____

8. $2x + y = 1$
    $x - y = 5$

8. _____

9. $3x - 2y = -4$
    $-3x + 2y = -2$

9. _____

10. $x + y = 5$
    $x - y = 1$

10. _____

11. $x - y = 10$
    $y = -2x + 5$

11. _____

12. $x = -3$
    $y = 4$

12. _____

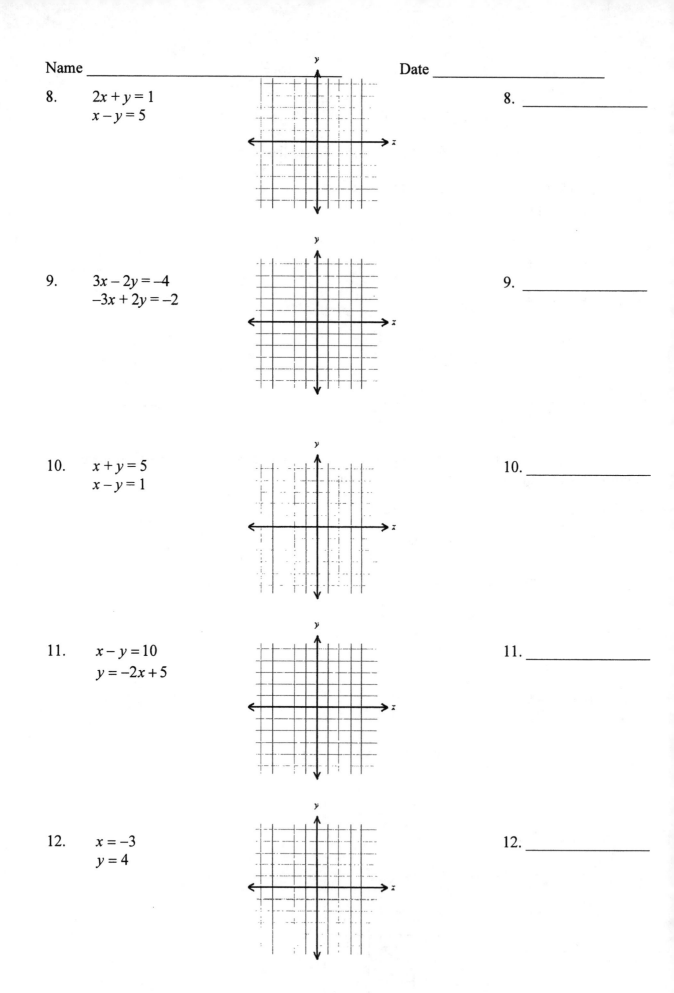

54

Name _____   Date _____

13. $2x + y = -2$
      $x - y = 5$

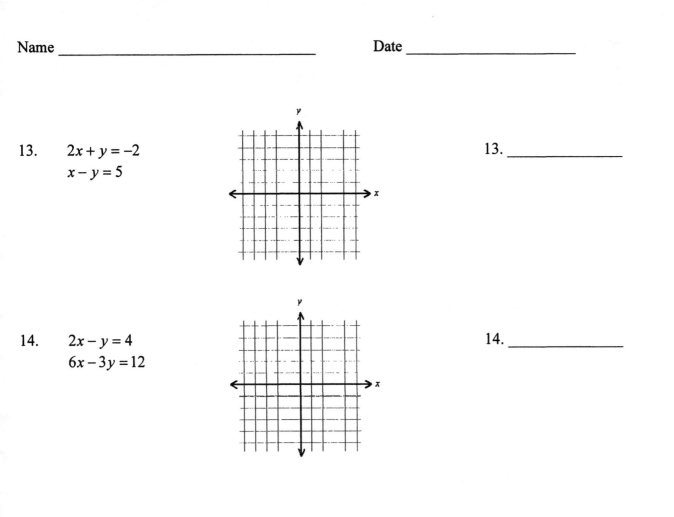

13. _____

14. $2x - y = 4$
      $6x - 3y = 12$

14. _____

Name _____   Date _____

Name _____   Date _____

Practice Set 5.2
Solving Systems of Linear Equations by the Substitution Method

Solve each system by the substitution method.

1. $x = 2y - 4$
    $3x + y = -5$

    1. _____

2. $3x - 3y = 12$
    $y = 2x - 8$

    2. _____

3. $x = 3y + 1$
    $4x + 7y = 0$

    3. _____

4. $y = x - 5$
    $2x + y = 4$

    4. _____

5. $x + y = 2$
    $x - y = 6$

    5. _____

6. $3x - 4y = -1$
    $4x - y = 3$

    6. _____

7. $4x + 2y = -3$
    $2x + y = 1$

    7. _____

8. $10x - 5y = 20$
    $x + 6y = -11$

    8. _____

9. $x + 3y = 5$
    $x - y = 1$

    9. _____

10. $3x + 3y = 5$
      $2x + 3y = 7$

10. _____

11. $4x - y = 10$
      $8x - 2y = 20$

11. _____

12. $3x - y = 7$
      $x + 2y = 7$

12. _____

13. $2x + y = 2$
      $0.4x + 0.5y = -2$

13. _____

14. $\dfrac{x}{3} + \dfrac{y}{4} = 1$
      $x + 5y = 3$

14. _____

15. $y = \dfrac{-3}{4}x + 5$

      $y = \dfrac{1}{2}x$

15. _____

16. $0.4x - 0.2y = 2.8$
      $y = x - 2$

16. _____

17. $5x - 5y = 10 + 2x$
      $2x + y = 14 - 3y$

17. _____

18. $6x + 8y = 12 - 2y$
      $2x + 5y = 12 - x$

18. _____

Practice Set 5.3
Solving Systems of Linear Equations by the Addition Method

Solve each system by the Addition Method. If there is no solution or infinitely many solutions, so state.

1. $x - y = 2$
           $-x + 2y = -8$
                    1. _____

2. $2x + y = 10$
           $3x - y = -5$
                    2. _____

3. $x + 4y = 3$
           $3x - 4y = -23$
                    3. _____

4. $2x - y = 6$
           $2x + y = 10$
                    4. _____

5. $x - y = -7$
           $3x + 2y = -6$
                    5. _____

6. $2x - 4y = 6$
           $x - y = -2$
                    6. _____

7. $3x + y = 1$
           $2x + y = 1$
                    7. _____

8. $3x + 6y = 12$
           $x + 2y = 4$
                    8. _____

9. $\dfrac{x}{6} + \dfrac{y}{4} = \dfrac{11}{12}$
           $\dfrac{x}{18} - \dfrac{y}{2} = \dfrac{-5}{18}$
                    9. _____

10. $2x - 4y = -1$
     $10x - 20y = 5$

10. _____

11. $3x - 2y = -9$
     $2x - 7y = 11$

11. _____

12. $5x + 11y = 14$
     $4x = 15y + 35$

12. _____

For the following six linear systems, solve two by graphing using the graphs provided at the end of the exercise. Solve two by the substitution method and two by the addition method. Evaluate the most efficient method for solving each system before beginning.

13. $y = -x$
     $y = -x + 4$

13. _____

14. $y = 5 - x$
     $3x - 4y = -20$

14. _____

15. $x + y = 10$
     $y = x + 8$

15. _____

16. $3x - 2y = 10$ 16. _____
        $5x + 3y = 4$

17. $y = 3$ 17. _____
        $x = 5$

18. $2x + y = 7$ 18. _____
        $x - y = 8$

Name _____   Date _____

Practice Set 5.4
Problem Solving Using Systems of Equations

Write a system of equations for each problem. Solve the system to solve the problem.

1. The sum of two numbers is 11. The first number is one less than twice the second. Find the two numbers.

    1. _____

2. The difference between two numbers is 14. When twice the first number is added to the second number, the result is 10. Find the two numbers.

    2. _____

3. The sum of two numbers is −16. One number is three times the other number. Find the two numbers.

    3. _____

4. Three hundred tickets were sold for the annual pancake breakfast. Adult tickets cost $5.00 and tickets for children under the age of twelve cost $2.50. Total receipts for the breakfast were $1,187.50. Find the number of adult tickets sold and the number of children's tickets sold.

    4. _____

5. The perimeter of a rectangle is 78 centimeters. The length of the rectangle is 6 centimeters less than twice the width. Find the dimensions of the rectangle.

    5. _____

6. The perimeter of a rectangle is 56 inches. The length of the rectangle is 4 inches less than three times the width. Find the dimensions of the rectangle.

    6. _____

7. The inmates of a state prison are responsible for maintaining a large vegetable garden. The garden is rectangular in shape and has a perimeter of 500 feet. The width is two-thirds as long as the length. Find the dimensions of the garden.

7. _____

8. Grace has 25 coins in her purse, all nickels and quarters. When she counted, she found she had a total of $3.85. How many nickels did she have? How many quarters?

8. _____

9. Heather bought 100 stamps at the post office. She bought some 30¢ stamps and some 39¢ stamps and spent a total of $34.95. How many of each kind of stamp did she buy?

9. _____

10. Maggie and Addison went shopping after Christmas. Maggie bought 4 pairs of socks and 2 sweaters, and spend $60. Addison spent $75 on 1 pair of socks and 3 sweaters. Find the cost of one pair of socks. Find the cost of one sweater.

10. _____

11. Jeff works two part-time jobs to make ends meet while attending college. He makes $6 an hour at the coffee shop and $8.45 an hour cleaning offices at night. If he worked 24 hours last week and made $180.75, how many hours did he work at each job?

11. _____

12. Amanda bought some fish for her new freshwater aquarium. She spend $45.80 on neon Tetras costing $2.00 each and giant Danios costing $3.40 each. If she bought 4 more neon Tetras than giant Danios, how many of each type fish did she buy?

12. _____

Name _____   Date _____

Practice Set 5.5
Systems of Linear Inequalities

Graph the solution of each system of linear inequality. To make the solution easier to see, use a different colored pencil to graph and shade each inequality in the system.

1. $x > 1$
    $y < 2$

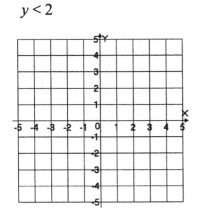

2. $y \geq x$
    $y < x + 1$

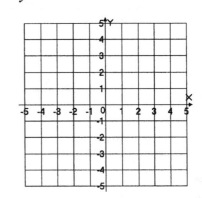

3. $y > 4x - 1$
    $y \leq -2x + 3$

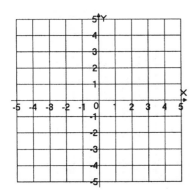

4. $y < 2x$
    $y \geq x - 5$

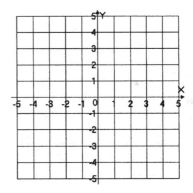

5. $x - 2y \geq 6$
    $x + 2y \leq 4$

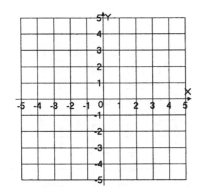

6. $2x + y > 8$
    $2x - y > 1$

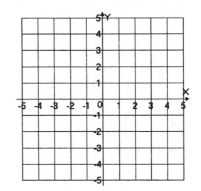

7. $x - y < -4$
        $2x + y > 0$

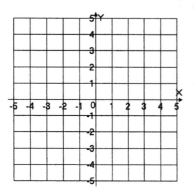

8. $y > 2x + 1$
        $x + y \geq -2$

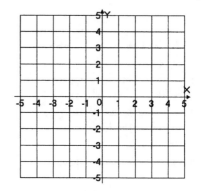

9. $x - y \leq 3$
        $x + 2y < 4$

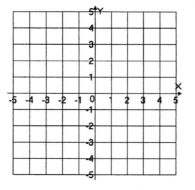

10. $y < -x + 5$
        $y > -x + 3$

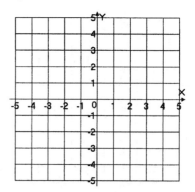

11. $y \leq -2x + 4$
        $3x - 5y \geq 15$

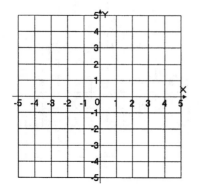

12. $x - y < 4$
        $x + y > 2$

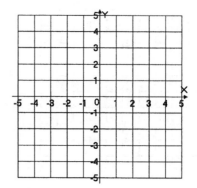

Practice Set 6.1
Adding and Subtracting Polynomials

Identify each polynomial as a monomial, a binomial, or a trinomial. Give the degree of the polynomial.

1. $4x - 9$ 1. _____

2. $x^4 - 3x^2 + 2$ 2. _____

3. $9x^3$ 3. _____

4. 5 4. _____

5. $x^2 + 4x - 7$ 5. _____

6. $-3x^{15}$ 6. _____

Add the following polynomials.

7. $\begin{aligned} 3x + 4 \\ \underline{-5x + 6} \end{aligned}$ 7. _____

8. $\begin{aligned} 5x^2 - 2x - 4 \\ \underline{3x^2 + 5x + 6} \end{aligned}$ 8. _____

9. $\begin{aligned} -8x^2 - 3x + 1 \\ \underline{-4x^2 + x + 8} \end{aligned}$ 9. _____

10. $\left(12x^3 - 4x^2 + 5x + 9\right) + \left(-2x^3 + 6x^2 - 3x - 8\right)$ 10. _____

11. $\left(2x^2 + 4x - 8\right) + \left(3x^3 - 4x - 2\right)$

11. _____

12. $\left(\dfrac{2}{3}x^2 + \dfrac{1}{5}x - \dfrac{3}{7}\right) + \left(-\dfrac{1}{3}x^2 + \dfrac{3}{5}x - \dfrac{3}{7}\right)$

12. _____

Subtract the following polynomials.

13. $9x + 4$
      $-(-2x + 5)$

13. _____

14. $7x^2 + 3x - 8$
      $-(-2x^2 - 4x + 6)$

14. _____

15. $3x^2 - 5x + 1$
      $-(4x^2 + 4x - 1)$

15. _____

16. $\left(0.5x^2 + 0.7x - 0.3\right) - \left(1.2x^2 - 2.4x + 1.1\right)$

16. _____

17. $\left(9x^3 - 7x\right) - \left(-2x^2 + 4x - 3\right)$

17. _____

18.   Subtract $3x^2 + 5x - 4$ from $-2x^2 + 7x - 8$

18. _____

Practice Set 6.2
Multiplying Polynomials

Multiply each expression using the product rule.

1. $x^2 \cdot x^3$ 1. _____

2. $y \cdot y^7$ 2. _____

3. $a^2 \cdot a \cdot a^4$ 3. _____

4. $5^3 \cdot 5^5$ 4. _____

Simplify each expression using the power rule.

5. $\left(3^4\right)^3$ 5. _____

6. $\left(x^7\right)^5$ 6. _____

7. $\left[(-10)^4\right]^4$ 7. _____

8. $\left[(-x)^5\right]^2$ 8. _____

Simplify each expression using the products-to-powers rule.

9. $(3x^2)^3$ 9. _____

10. $(-5x^3)^2$ 10. _____

11. $(-2y^3)^4$ 11. _____

12. $(-2x^9)^3$ 12. _____

Multiply the monomials.

13. $(5x^2)(6x^3)$ 13._____

14. $(-3a^2)(-3a^5)$ 14._____

15. $(8x^2)(2x^3)$ 15._____

16. $(3y^3)(-5y^4)$ 16._____

Find each product.

17. $4x^2(3x^2 - 4x + 5)$ 17._____

18. $(x^3 + 2x^2 - x)(-3x)$ 18._____

19. $(3x - 2)(4x + 1)$ 19._____

20. $(5x + 3)(5x - 3)$ 20._____

21. $(2x + 7)(3x - 1)$ 21._____

22. $(5x + 4)(2x - 3)$ 22._____

23. $(4x + 1)(x - 7)$ 23._____

24. $(x - 1)(x^2 - 2x + 4)$ 24._____

25. $(x^2 - 3x + 1)(x^2 + 3x - 1)$ 25._____

Name _____   Date _____

Practice Set 6.3
Special Products

Use the FOIL method to find each product. Express the product in descending powers of the variable.

1. $(x+7)(x+3)$ 1. _____

2. $(2x+3)(3x-1)$ 2. _____

3. $(5x+4)(3x-2)$ 3. _____

4. $(2x+1)(3x-8)$ 4. _____

5. $(7x^2-5)(4x^2+2)$ 5. _____

Multiply using the rule for finding the product of the sum and difference of two terms.

6. $(x+4)(x-4)$ 6. _____

7. $(9x+1)(9x-1)$ 7. _____

8. $(3x+5)(3x-5)$ 8. _____

9. $\left(4x+\dfrac{1}{4}\right)\left(4x-\dfrac{1}{4}\right)$ 9. _____

10. $(x^2+6)(x^2-6)$ 10. _____

Multiply using the rules for the square of a binomial.

11. $(2x+5)^2$ 11. _____

12. $(7x+3)^2$ 12. _____

13. $(x-1)^2$ 13. _____

14. $(4-5x)^2$ 14. _____

15. $\left(3y-\dfrac{1}{3}\right)^2$ 15. _____

Multiply using the method of your choice.

16. $9x^2(4x^2+x-3)$ 16. _____

17. $(x+2)^2$ 17. _____

18. $(x^2+3)^2$ 18. _____

19. $(3y-2)(3y+2)$ 19. _____

20. $(x-3)(x^2+3x+9)$ 20. _____

Practice Set 6.4
Polynomials in Several Variables

Evaluate each polynomial for $x = 2$ and $y = -3$.

1.　　$x^2 - 3xy + y^2$　　　　　　　　　　　　　　1. _____

2.　　$x^3 y - 5xy + 1$　　　　　　　　　　　　　　2. _____

In exercises 3-5, (a) determine the coefficient of each term, (b) the degree of each term, and (c) the degree of the polynomial.

3.　　$4a^3 b^2 - 3a^2 b + 4ab^5$　　　　　　　　　3a. _____

　　　　　　　　　　　　　　　　　　　　　　　　　　b. _____

　　　　　　　　　　　　　　　　　　　　　　　　　　c. _____

4.　　$5a^4 b^3 + 2a^3 b - a^2 b^3 + 3ab^5 + b^6$　　4a. _____

　　　　　　　　　　　　　　　　　　　　　　　　　　b. _____

　　　　　　　　　　　　　　　　　　　　　　　　　　c. _____

5.　　$13x^5 + 3xy^6 - 5x^2 y^5 + x^3 y^7$　　　　5a. _____

　　　　　　　　　　　　　　　　　　　　　　　　　　b. _____

　　　　　　　　　　　　　　　　　　　　　　　　　　c. _____

Perform the indicated operations.

6.　　$\left(6x^2 - 3xy + 4y^2\right) + \left(5x^2 - 6xy - 3y^2\right)$　　6. _____

7. $(-3x^2y + xy + 6) + (2x^2y - 5xy - 11)$ 7. _____

8.   Add $(5x^2y^2 + 3xy^2 + 7y^2)$ and $(-6x^2y^2 - 4xy^2 - 2y^2)$ 8. _____

9. $\left(14a^3 + 3a^2b - 5ab^3 + 3b^4\right) - \left(8a^3 - 5a^2b + 2ab^3 - 3b^4\right)$ 9. _____

10. $(x^3 - y^3) - (-2x^3 - 2x^2y + 3xy^2 + 4y^3)$ 10. _____

11.   Subtract $(2a^2b^4 - 5ab^2 + 3ab)$ from $(3a^2b^4 + 3ab^2 + ab)$ 11. _____

12. $(4x^2y)(3xy)$ 12. _____

13. $(-5x^3y^4)(-6x^4y^2)$ 13. _____

14. $3ab^2(6a^3b^2 - 4ab)$ 14. _____

15. $\left(3x - 4y\right)^2$ 15. _____

16. $\left(8x + 5y\right)\left(2x - y\right)$ 16. _____

17. $4x^3y\left(3x^2 - 2xy + 5x^3y^4\right)$ 17. _____

18. $\left(7x^2 - 3y\right)\left(x^2 + y\right)$ 18. _____

19. $\left(2x + 5y\right)\left(2x - 5y\right)$ 19. _____

20. $(x + y + 3)(x - y - 3)$ 20. _____

Name _____   Date _____

Divide each expression using the quotient rule. Express any numerical answers in exponential form.

1. $\dfrac{5^7}{5^2}$ 1. _____

2. $\dfrac{x^9}{x^6}$ 2. _____

3. $\dfrac{x^5 y^{10}}{x^2 y}$ 3. _____

4. $\dfrac{x^{20} y^{17}}{x^9 y^4}$ 4. _____

Use the zero-exponent rule to simplify each expression.

5. 3^0 5. _____

6. $(-4)^0$ 6. _____

7. -5^0 7. _____

8. $45x^0$ 8. _____

Simplify each expression using the quotients-to-powers rule.

9. $\left(\dfrac{x}{4}\right)^2$ 9. _____

10. $\left(\dfrac{2x^3}{3}\right)^3$ 10. _____

11. $\left(\dfrac{-6x^4}{7}\right)^2$ 11. _____

12. $\left(\dfrac{x^3 y^4}{2z^2}\right)^4$ 12. _____

Divide the monomials.

13. $\dfrac{25x^9}{5x^2}$ 13. _____

14. $\dfrac{-5x^{10} y^7 z^3}{10x^5 y^2 z}$ 14. _____

15. $\dfrac{30a^2 b^4 c^8}{5abc}$ 15. _____

16. $\dfrac{-20x^{12} y^9}{35x^7 y^5}$ 16. _____

Divide the polynomial by the monomial.

17. $\dfrac{40x^5 + 15x^2}{5}$ 17. _____

18. $\dfrac{15y^4 - 42y^3}{12y^2}$ 18. _____

19. $\dfrac{12x^5 y^6 - 18x^3 y^7 + 6x^2 y^8}{-3x^2 y^4}$ 19. _____

20. $\dfrac{20x^5 + 15x^3 + 5x^2}{5x^2}$ 20. _____

Practice Set 6.6
Dividing Polynomials by Binomials

Divide as indicated.

1. $(x^2 + 3x - 10) \div (x + 5)$

1. _____

2. $(x^3 - 27) \div (x - 3)$

2. _____

3. $(x^2 - 2x) \div (x + 2)$

3. _____

4. $\left(6x^2 + 2x - 28\right) \div \left(3x + 7\right)$

4. _____

5. $\left(4x^2 - 6 + 4x^3 - 7x\right) \div \left(2x + 3\right)$

5. _____

6. $\left(6x^3 + 11x^2 - 4x - 9\right) \div \left(3x - 2\right)$

6. _____

7. $\left(8x^3 + 30x^2 - 36x - 12\right) \div \left(4x - 1\right)$

7. _____

8. $\left(x^3 - 13x - 12\right) \div \left(x - 4\right)$

8. _____

9. $(3x^4 + 4 - 10x + 7x^3) \div \left(3x - 2\right)$

9. _____

10. $(3x^3 + 20x^2 + 21) \div (x + 7)$

10. _____

Name _____   Date _____

78

Practice Set 6.7
Negative Exponents and Scientific Notation

Write each expression with positive exponents only. Then simplify, if possible.

1. 3^{-3} 1. _____

2. $(-6)^{-2}$ 2. _____

3. $3^{-1} + 4^{-1}$ 3. _____

4. $\dfrac{1}{2x^{-4}}$ 4. _____

Simplify each exponential expression. Assume that variables represent non-zero real numbers.

5. $x^{12} \cdot x^{-5}$ 5. _____

6. $\dfrac{-3a^7}{18a^9}$ 6. _____

7. $\dfrac{x^5}{(x^2)^{-3}}$ 7. _____

8. $(3x^{-2})(2x^5)$ 8. _____

Write each number in decimal notation without the use of exponents.

9. 3.1×10^4 9. _____

10. 6.017×10^0 10. _____

11. 5.32×10^{-2}

11. _____

12. 9.146×10^{-3}

12. _____

Write each number in scientific notation.

13.   15,800

13. _____

14.   9,281

14. _____

15.   0.0057

15. _____

16.   0.000114

16. _____

Perform the indicated computations. Write the answers in scientific notation.

17. $(2 \times 10^3)(2 \times 10^4)$

17. _____

18. $(4 \times 10^6)(2 \times 10)$

18. _____

19. $\dfrac{8 \times 10^6}{2 \times 10^{-3}}$

19. _____

20. $\dfrac{4 \times 10^5}{1 \times 10^{-2}}$

20. _____

Name _____   Date _____

Practice Set 7.1
The Greatest Common Factor and Factoring by Grouping

Factor each polynomial by factoring out the greatest common factor. If there is no common factor other than 1, and the polynomial cannot be factored, so state.

1. $9x + 9$ 1. _____

2. $7x + 21$ 2. _____

3. $x^3 + 4x^2$ 3. _____

4. $8x^3 + 36x^2$ 4. _____

5. $18x^4 + 9x^2 + 3$ 5. _____

6. $15x^5 + 7x^2$ 6. _____

7. $4x^3 + 9$ 7. _____

8. $20x^4 + 16x^2 - 8x$ 8. _____

9. $6x^2y^3 + 15x^4y^2$ 9. _____

10. $12x^3y^4 + 30x^2y^5 + 48xy^6$ 10. _____

Factor each polynomial by factoring out the greatest common binomial factor.

11. $x(x-4)+3(x-4)$ 11._____

12. $2x(x+6)+5(x+6)$ 12._____

13. $7x(x+9)-(x+9)$ 13._____

Factor each polynomial by grouping.

14. $xy-4x+2y-8$ 14._____

15. $14+2a-7b-ab$ 15._____

16. $bx+5x-2by-10y$ 16._____

17. $mn+m+8n+8$ 17._____

18. $14-7y-2x+xy$ 18._____

19. $2a+6+3b+ab$ 19._____

20. $xy-2x-5y+10$ 20._____

Name _____   Date _____

Practice Set 7.2
Factoring Trinomials Whose Leading Coefficient is 1

Factor completely.

1. $x^2 + 6x + 8$

1. _____

2. $y^2 + 8y + 15$

2. _____

3. $x^2 + 3x - 4$

3. _____

4. $m^2 + 2m - 99$

4. _____

5. $y^2 - 8y + 12$

5. _____

6. $y^2 - 13y + 36$

6. _____

7. $a^2 - 5a - 14$

7. _____

8. $x^2 + 4xy + 3y^2$

8. _____

9. $m^2 - 10m + 24$

9. _____

10. $x^2 - 2x - 24$

10. _____

11. $5x^2 + 25x + 20$ 11. _____

12. $a^2 + 9a + 8$ 12. _____

13. $x^2 - 8x + 7$ 13. _____

14. $7x^2 - 35x + 42$ 14. _____

15. $x^2 + 3xy - 10y^2$ 15. _____

16. $8a^2 + 16a - 24$ 16. _____

17. $2y^3 + 16y^2 + 24y$ 17. _____

18. $m^2 + 9m + 20$ 18. _____

19. $x^2 - 15x + 56$ 19. _____

20. $6x^3 + 54x^2 + 108x$ 20. _____

Practice Set 7.3
Factoring Trinomials Whose Leading Coefficient is Not 1

Factor each trinomial completely by the method of your choice. If the trinomial will not factor, state that it is prime. Be sure to check your factorization.

1. $9x^2 + 15x + 4$ 1. _____

2. $12x^2 - 8x - 15$ 2. _____

3. $4x^2 - 16x + 15$ 3. _____

4. $8y^2 + 6y + 1$ 4. _____

5. $4a^2 - 12a + 9$ 5. _____

6. $20x^2 + 53x + 18$ 6. _____

7. $2x^2 + 13x + 15$ 7. _____

8. $3y^2 + 14y + 8$ 8. _____

9. $2x^2 + 5x - 7$ 9. _____

10. $4x^2 - 13x + 10$ 10. _____

11. $10x^2 + 19x + 6$ 11. _____

12. $24x^2 - 2x + 2$ 12. _____

13. $18a^2 - 3a - 10$ 13. _____

14. $10x^2 - 29xy + 10y^2$ 14. _____

15. $16y^2 - 8x - 15$ 15. _____

16. $12x^2 - 25xy + 12y^2$ 16. _____

17. $40a^2 + 58a - 21$ 17. _____

18. $20y^4 + 29y^3 - 36y^2$ 18. _____

19. $3a^3 - 18a^2 - 48a$ 19. _____

20. $8x^2 + 22xy + 5y^2$ 20. _____

Practice Set 7.4
Factoring Special Forms

Factor completely or state that the polynomial is prime.

1. $x^2 - 16$ 1. _____

2. $4y^2 - 49$ 2. _____

3. $5x^2 - y^2$ 3. _____

4. $25a^2 - 16$ 4. _____

5. $x^2 + 16x + 64$ 5. _____

6. $x^2 - 3x + 1$ 6. _____

7. $12x^2 - 3y^2$ 7. _____

8. $x^2 + 12x + 36$ 8. _____

9. $4x^2 - 12xy + 9y^2$ 9. _____

10. $x^4 - 81$ 10. _____

11. $25y^2 + 10y + 1$ 11. _____

12. $a^3 - 8b^3$ 12. _____

13. $x^2 - 1$ 13. _____

14. $27x^3 + 64$ 14. _____

15. $y^2 - 14y + 49$ 15. _____

16. $8x^3 + 1$ 16. _____

17. $9m^2 - 48m + 64$ 17. _____

18. $3y^3 - 81$ 18. _____

19. $x^6 - 25y^2$ 19. _____

20. $4x^2 + 44x + 121$ 20. _____

Practice Set 7.5
A General Factoring Strategy

Factor completely, or state that the polynomial is prime.

1. $x^2 - 196$ 1. _____

2. $a^2 + 12a + 36$ 2. _____

3. $y^2 - 5y - 14$ 3. _____

4. $6x^2 - 51x - 90$ 4. _____

5. $18m^2 - 50$ 5. _____

6. $4x^2 + 49y^2$ 6. _____

7. $x^2 + 3ax - 2bx - 6ab$ 7. _____

8. $5x^2 + 14x - 3$ 8. _____

9. $x^3 + 64y^3$ 9. _____

10. $2x^3 + 4x^2 - 30x$ 10. _____

11. $x^3 + 5x^2 - 16x - 80$ 11. _____

12. $4a^2 - 16a + 16$ 12. _____

13. $2x^2 - 32$ 13. _____

14. $y^3 + 4y^2 - 9y - 36$ 14. _____

15. $20x^2 + 29xy + 5y^2$ 15. _____

16. $3a^2 + 18a + 27$ 16. _____

17. $8a^3 + 125$ 17. _____

18. $xy - 2x + 5y - 10$ 18. _____

19. $6x^2 + x - 35$ 19. _____

20. $x^2 + 5x + 9$ 20. _____

Name _____   Date _____

Practice Set 7.6
Solving Quadratic Equations by Factoring

Solve each equation using the zero-product principle.

1. $a(a+4) = 0$

    1. _____

2. $2x(x-5) = 0$

    2. _____

3. $(y-3)(y+7) = 0$

    3. _____

4. $(2x-1)(3x+4) = 0$

    4. _____

Use factoring to solve each quadratic equation.

5. $x^2 + 7x + 12 = 0$

    5. _____

6. $x^2 + 3x - 10 = 0$

    6. _____

7. $x^2 - x = 20$

    7. _____

8. $x^2 - 16 = 0$

    8. _____

9. $(x+2)(x-5) = 8$

    9. _____

10. $10a^2 = 3a + 4$

    10. _____

11. $7x^2 = 28x$

11. _____

12. $x(x+6) = 40$

12. _____

13. $(x+1)(x+2) = 12$

13. _____

14. $3x^2 - 5x + 2 = 0$

14. _____

15. $8x^2 + 2x = 1$

15. _____

16. $(x+1)^2 = 3x + 7$

16. _____

17. $(x+6)(x-2) = -7$

17. _____

Solve each problem.

18. The product of two positive numbers is 150. One number is 5 less than twice the other number. Find the two numbers.

18. _____

19. The area of a rectangle is 32 square inches. The length is 4 more than the width. Find the dimensions of the rectangle.

19. _____

20. One number is 3 more than another number. The sum of their squares is 29. Find the two numbers.

20. _____

Name _____   Date _____

Practice Set 8.1
Rational Expressions and Their Simplification

Find all numbers for which the rational expression is undefined. If the rational expression is defined for all real numbers, so state.

1. $\dfrac{7}{4x}$

1. _____

2. $\dfrac{19}{3x-9}$

2. _____

3. $\dfrac{x+8}{8}$

3. _____

4. $\dfrac{11}{x^2+9}$

4. _____

Simplify each rational expression.

5. $\dfrac{6x+12}{3}$

5. _____

6. $\dfrac{4x}{12x-16}$

6. _____

7. $\dfrac{3a+6}{a+2}$

7. _____

8. $\dfrac{a^2-9}{a^2+5a+6}$

8. _____

9. $\dfrac{2t^2+6t+4}{4t^2-12t-16}$

9. _____

10. $\dfrac{a^2-6a+9}{a^2-9}$

10. _____

11. $\dfrac{-5y-5}{y^2+7y+6}$ 11. _____

12. $\dfrac{x^2-25}{x^2-10x+25}$ 12. _____

13. $\dfrac{y^2+4}{y+2}$ 13. _____

14. $\dfrac{x-5}{x^2-3x-10}$ 14. _____

15. $\dfrac{x+1}{x^2-1}$ 15. _____

16. $\dfrac{x-5}{5-x}$ 16. _____

17. $\dfrac{x^3-27}{x-3}$ 17. _____

18. $\dfrac{x+4}{4}$ 18. _____

19. $\dfrac{4x+20}{x^2+6x+5}$ 19. _____

20. $\dfrac{x^2-x-2}{2-x}$ 20. _____

Name _____   Date _____

Practice Set 8.2
Multiplying and Dividing Rational Expressions

Multiply as indicated.

1. $\dfrac{3}{x+2} \cdot \dfrac{x-4}{5}$

   1. _____

2. $\dfrac{2}{x} \cdot \dfrac{3x}{7}$

   2. _____

3. $\dfrac{a-2}{a^2-1} \cdot \dfrac{a-1}{a^2-5a+6}$

   3. _____

4. $\dfrac{y^2-2y+1}{5y} \cdot \dfrac{10y^2}{y-1}$

   4. _____

5. $\dfrac{x+3}{x-3} \cdot \dfrac{x^2-6x+9}{2x+6}$

   5. _____

6. $\dfrac{x^2+7x+12}{x^2+2x-8} \cdot \dfrac{x^2+3x-10}{x^2+8x+15}$

   6. _____

7. $\dfrac{x^3-27}{x^2-9} \cdot \dfrac{x+3}{x-3}$

   7. _____

8. $\dfrac{x^2+x-2}{x^2+5x+6} \cdot \dfrac{x^2+2x-3}{x^2+2x+1}$

   8. _____

9. $\dfrac{xy+4y-2x-8}{x^2-16} \cdot \dfrac{5x-20}{y^2+y-6}$

   9. _____

10. $\dfrac{x^2-5x+6}{2x+4} \cdot \dfrac{x+2}{2x-6}$

    10. _____

Divide as indicated.

11. $\dfrac{x}{4} \div \dfrac{9}{5}$

11. _____

12. $\dfrac{x+3}{5} \div \dfrac{2x+6}{7}$

12. _____

13. $\dfrac{x^2-4x}{9} \div \dfrac{x-4}{6}$

13. _____

14. $\dfrac{6}{x^2-1} \div \dfrac{3}{x+1}$

14. _____

15. $\dfrac{y^2-4y+4}{y+2} \div (y-2)$

15. _____

16. $\dfrac{x^2-3x}{3x+7} \div \dfrac{2x^3-6x^2}{9x+21}$

16. _____

17. $\dfrac{6a+30}{7a^2-11a-6} \div \dfrac{3a-3}{7a^2-4a-3}$

17. _____

18. $\dfrac{x^2-3x-4}{4x^2+12x} \div \dfrac{x^2-2x-8}{x^2+6x+8}$

18. _____

19. $\dfrac{x^2+7x+2}{3x-3} \div \dfrac{x^2+7x+2}{x-1}$

19. _____

20. $\dfrac{x^2+x-2}{x^2+5x+6} \div \dfrac{x-1}{x}$

20. _____

Practice Set 8.3
Adding and Subtracting Rational Expressions with the Same Denominator

Add or subtract as indicated. Simplify, if possible.

1. $\dfrac{4x}{7} + \dfrac{2x}{7}$

1. _____

2. $\dfrac{x+2}{15} + \dfrac{4x-8}{15}$

2. _____

3. $\dfrac{x}{4} - \dfrac{3x-5}{4}$

3. _____

4. $\dfrac{4x+5}{x-1} - \dfrac{2x-1}{x-1}$

4. _____

5. $\dfrac{x^2-10x-7}{2x+1} + \dfrac{x^2+x+2}{2x+1}$

5. _____

6. $\dfrac{3x}{x+7} - \dfrac{x-2}{x+7}$

6. _____

7. $\dfrac{2x^2+3x-7}{2x+1} + \dfrac{x^2+x-8}{2x+1}$

7. _____

8. $\dfrac{4x^3-2x}{8x^2} - \dfrac{6x^3+4x}{8x^2}$

8. _____

9. $\dfrac{x+8}{x+7} + \dfrac{10-4x}{x+7}$

9. _____

Name _____   Date _____

In exercises 10-15, the denominators are opposites or additive inverses. Add or subtract as indicated. Simplify the result, if possible.

10. $\dfrac{2}{x-3} + \dfrac{2}{3-x}$

10. _____

11. $\dfrac{2x+1}{x-3} - \dfrac{x+2}{3-x}$

11, _____

12. $\dfrac{y^2}{y-7} - \dfrac{49}{7-y}$

12. _____

13. $\dfrac{3x}{x^2-y^2} + \dfrac{3y}{y^2-x^2}$

13. _____

14. $\dfrac{x-2}{x^2-4} - \dfrac{x-2}{4-x^2}$

14. _____

15. $\dfrac{2x+4}{x^2+x-30} + \dfrac{x-2}{30-x-x^2}$

15. _____

.

Name _____   Date _____

Adding and Subtracting Rational Expressions with Different Denominators

Find the least common denominator of the rational expressions.

1. $\dfrac{4}{8x}$ and $\dfrac{12}{10x^3}$ 1. _____

2. $\dfrac{6x}{x+4}$ and $\dfrac{5}{x-3}$ 2. _____

3. $\dfrac{x+2}{5x^2+10x}$ and $\dfrac{13}{x^2-4}$ 3. _____

4. $\dfrac{5y}{y^2-y-12}$ and $\dfrac{2}{2y^2-3y-9}$ 4. _____

Add or subtract as indicated. Simplify the result, if possible.

5. $\dfrac{2x+3}{3x}-\dfrac{3-2x}{3x}$ 5. _____

6. $5+\dfrac{1}{x}$ 6. _____

7. $\dfrac{x-3}{2}+\dfrac{x+2}{4}$ 7. _____

8. $\dfrac{8}{x^2-16}-\dfrac{7}{x^2-x-12}$ 8. _____

9. $\dfrac{5}{y}+\dfrac{20}{y^2-4y}$ 9. _____

Name _____   Date _____

10. $\dfrac{3}{2x+10}+\dfrac{8}{3x+15}$ 10. _____

11. $\dfrac{y}{y+3}+\dfrac{5}{y^2-9}$ 11. _____

12. $\dfrac{a-8}{a^2-a-2}+\dfrac{2}{a-2}$ 12. _____

13. $\dfrac{6}{x^2+x-20}-\dfrac{5}{x^2+8x+15}$ 13. _____

14. $\dfrac{x^2+4x-5}{x^2-2x-3}-\dfrac{2}{x+1}$ 14. _____

15. $\dfrac{y+1}{y-1}-\dfrac{4}{y^2-1}$ 15. _____

Name _____   Date _____

Practice Set 8.5
Complex Rational Expressions

Simplify each complex rational expression by the method of your choice.

1. $\dfrac{\dfrac{5}{6}+\dfrac{2}{3}}{\dfrac{1}{2}+\dfrac{1}{6}}$

1. _____

2. $\dfrac{\dfrac{1}{2}+\dfrac{5}{16}}{\dfrac{3}{4}+\dfrac{3}{8}}$

2. _____

3. $\dfrac{\dfrac{5}{8}+\dfrac{2}{x}}{\dfrac{1}{4}+\dfrac{3}{x}}$

3. _____

4. $\dfrac{1+\dfrac{3}{y}}{1-\dfrac{9}{y^2}}$

4. _____

5. $\dfrac{x+\dfrac{6}{y}}{\dfrac{x}{y}}$

5. _____

6. $\dfrac{2+\dfrac{4}{y}}{3-\dfrac{1}{y}}$

6. _____

7. $\dfrac{\dfrac{1}{x} + \dfrac{3}{x^2}}{\dfrac{2}{x} + 1}$

7. _____

8. $\dfrac{\dfrac{1}{x} - \dfrac{2}{y}}{\dfrac{1}{x} + \dfrac{2}{y}}$

8. _____

9. $\dfrac{\dfrac{2}{x+2} - \dfrac{2}{x-2}}{\dfrac{5}{x^2 - 4}}$

9. _____

10. $\dfrac{\dfrac{1}{x} - \dfrac{2}{x^2}}{\dfrac{2}{x} - 1}$

10. _____

11. $\dfrac{\dfrac{1}{a^2} + \dfrac{2}{ab}}{\dfrac{3}{a^2 b} + \dfrac{1}{b}}$

11. _____

12. $\dfrac{\dfrac{2}{x} + \dfrac{x}{2}}{\dfrac{y}{3} - \dfrac{3}{y}}$

12. _____

13. $\dfrac{x - \dfrac{3}{y}}{\dfrac{1}{x} - \dfrac{8}{y}}$

13. _____

14. $\dfrac{\dfrac{1}{x+1} + 2}{\dfrac{1}{x+1} + 3}$

14. _____

15. $\dfrac{1 - \dfrac{1}{x-1}}{3 + \dfrac{1}{x+1}}$

15. _____

16. $\dfrac{x + \dfrac{27}{x^2}}{1 - \dfrac{3}{x} + \dfrac{9}{x^2}}$

16. _____

Name _____ Date _____

Practice Set 8.6
Solving Rational Equations

Solve each rational equation.

1. $\dfrac{x}{4} = \dfrac{x}{3} - 6$

1. _____

2. $\dfrac{2}{x} + 4 = \dfrac{3}{2x} + \dfrac{9}{4}$

2. _____

3. $\dfrac{a}{a-3} = \dfrac{3}{2}$

3. _____

4. $\dfrac{5}{x+1} = \dfrac{4}{x+2}$

4. _____

5. $\dfrac{5}{x} - \dfrac{1}{3} = \dfrac{3}{x}$

5. _____

6. $1 + \dfrac{12}{y} = \dfrac{-35}{y^2}$

6. _____

7. $\dfrac{50}{x+5} = \dfrac{30}{x-5}$

7. _____

8. $\dfrac{10}{x+4} = \dfrac{6}{x} - \dfrac{4}{x}$

8. _____

9. $\dfrac{10}{x^2-25} = \dfrac{3}{x+5} + \dfrac{1}{x-5}$

9. _____

10. $\dfrac{3}{x+6} - \dfrac{1}{x-2} = \dfrac{-8}{x^2+4x-12}$

10. _____

11. $\dfrac{2a}{a-3} + \dfrac{6-2a}{a^2-9} = \dfrac{a}{a+3}$

11. _____

12. $\dfrac{x+2}{x^2+3x} = \dfrac{-2}{x^2-9}$

12. _____

13. $\dfrac{5}{x^2-9} = \dfrac{1}{x-3} - \dfrac{2}{x+3}$

13. _____

14. $\dfrac{6}{x^2-2x-3} = \dfrac{2}{x^2-1}$

14. _____

Practice Set 8.7
Applications Using Rational Equations and Variation

1. In still water, a boat averages 25 miles per hour. It takes the same amount of time to travel 6 miles downstream with the current as it takes to travel 4 miles upstream against the current. What is the rate of the water's current?

1. _____

2. The water's current is 4 miles per hour. A boat can travel 30 miles downstream, with the current, in the same time it takes to travel 22 miles upstream, against the current. What is the boat's average rate in still water?

2. _____

3. In still water, a boat averages 16 miles per hour. It takes the same amount of time to travel 26 miles up the river against the current as 38 miles down the river with the current. What is the rate of the current?

3. _____

4. Coleman can grade a set of test papers in 15 minutes. Stacy can grade the same number of test papers in 25 minutes. How long will it take them to grade the set of papers working together?

4. _____

5. There are two separate drains to the tank at the farm. When the larger drain is used alone, it can empty the tank in 5 hours. If the smaller of the two drains is used alone, it will take 6 hours. How long will it take to empty the tank if both drains are used?

5. _____

6. Becky can wash her car in 30 minutes. Her sister, Glenda can wash the car alone in 45 minutes. If they work together, how long will it take them to wash the car?

6. _____

Name _____   Date _____

7. If a six-foot lamp post casts a 14 foot shadow, how tall would   7. _____
    a flag pole be if it casts a 35 foot shadow?

8. Triangle PQR is similar to triangle XYZ. $\overline{PQ}$ measures   8. _____
    24 inches, $\overline{QR}$ measures 8 inches, $\overline{XY}$ measures 3 inches.
    Find the measure of $\overline{YZ}$.

9. If y varies directly as x and $y = 18$ when $x = 6$, find y if $x = 20$.   9. _____

10. If y varies inversely as x and $y = 7$ when $x = 10$, find y if $x = 5$.   10. _____

11. If y varies indirectly as x and $y = 4$ when $x = 8$, find y if $x = 16$.   11. _____

12. If y varies inversely as x and $y = 9$ when $x = 2$, find y if $x = 54$.   12. _____

Practice Set 9.1
Finding Roots

Evaluate each expression, or state that the expression is not a real number.

1. $\sqrt{64}$ 1. _____

2. $\sqrt{-9}$ 2. _____

3. $-\sqrt{25}$ 3. _____

4. $\sqrt[3]{-27}$ 4. _____

5. $-\sqrt{64+36}$ 5. _____

6. $\sqrt{64}+\sqrt{36}$ 6. _____

7. $\sqrt{\dfrac{49}{100}}$ 7. _____

8. $\sqrt[3]{125}$ 8. _____

9. $\sqrt[5]{-1}$ 9. _____

10. $\sqrt{\dfrac{1}{36}}$ 10. _____

11. $\sqrt[3]{64}$

11. _____

12. $\sqrt[4]{16}$

12. _____

13. $\sqrt{51-60}$

13. _____

14. $\sqrt{\dfrac{100}{4}}$

14. _____

15. $\sqrt[3]{-1000}$

15. _____

Use a calculator to approximate each expression. Round to three decimal places. If the expression is not a real number, so state.

16. $\sqrt{14}$

16. _____

17. $\sqrt{80}$

17. _____

18. $\sqrt{10}$

18. _____

19. $\sqrt{7-15}$

19. _____

20. $-\sqrt{24}$

20. _____

Practice Set 9.2
Multiplying and Dividing Radicals

Simplify each radical expression. Assume that variable expressions in radicands represent positive real numbers.

1. $\sqrt{18}$ 1. _____

2. $\sqrt{45}$ 2. _____

3. $\sqrt{20x^2}$ 3. _____

4. $\sqrt{56y^2}$ 4. _____

5. $\sqrt{72a^8}$ 5. _____

6. $\sqrt{44x^6}$ 6. _____

7. $\sqrt[3]{54y^3}$ 7. _____

8. $\sqrt{\dfrac{3}{4}}$ 8. _____

9. $\sqrt{28y^{12}}$ 9. _____

10. $\sqrt[3]{\dfrac{3}{8}}$ 10. _____

11. $\sqrt{300x^{15}}$

11. _____

12. $\dfrac{\sqrt{50x^9}}{\sqrt{5x^3}}$

12. _____

13. $\sqrt{108a^5}$

13. _____

14. $\dfrac{\sqrt{40x^2}}{\sqrt{10x^2}}$

14. _____

15. $\sqrt{\dfrac{100x^9}{5x^2}}$

15. _____

Multiply and simplify, if possible.

16. $\sqrt{5} \cdot \sqrt{15}$

16. _____

17. $\sqrt{3x} \cdot \sqrt{3x}$

17. _____

18. $\sqrt{6a^3} \cdot \sqrt{10a^3}$

18. _____

19. $\sqrt[4]{4y^3} \cdot \sqrt[4]{8y}$

19. _____

20. $\sqrt[3]{40x} \cdot \sqrt[3]{2x^6}$

20. _____

Practice Set 9.3
Operations with Radicals

Add or subtract as indicated. If the terms are not like radicals and cannot be combined, so state.

1. $2\sqrt{6} + 7\sqrt{6}$

    1. _____

2. $5\sqrt{x} - 4\sqrt{x}$

    2. _____

3. $2\sqrt{5} + 3\sqrt{5} + \sqrt{5}$

    3. _____

4. $\sqrt{3} + 4\sqrt{2}$

    4. _____

5. $\sqrt{27} + \sqrt{48}$

    5. _____

6. $8\sqrt{3} - 2\sqrt{3}$

    6. _____

7. $7\sqrt{3y} + 8\sqrt{3y}$

    7. _____

8. $9\sqrt{12} + 4\sqrt{27}$

    8. _____

9. $\sqrt{32} - \sqrt{2}$

    9. _____

10. $5\sqrt{10} + 3\sqrt{10} - \sqrt{10}$

    10. _____

Name _____   Date _____

Multiply as indicated. Simplify radicals as needed.

11. $\sqrt{6}(\sqrt{5}+2)$ 11. _____

12. $\sqrt{7}(\sqrt{2}-\sqrt{6})$ 12. _____

13. $\sqrt{10}(\sqrt{2}+\sqrt{3})$ 13. _____

14. $(3\sqrt{5}+4)(3\sqrt{5}-4)$ 14. _____

15. $(\sqrt{x}+\sqrt{2})^2$ 15. _____

16. $(\sqrt{3}+2)(\sqrt{3}-4)$ 16. _____

17. $(\sqrt{2}+3\sqrt{2})(2\sqrt{2}+4\sqrt{2})$ 17. _____

18. $(3\sqrt{5}-1)(2\sqrt{5}+1)$ 18. _____

19. $(4\sqrt{6}+2)(2\sqrt{3}-5)$ 19. _____

20. $(\sqrt{x}-3)^2$ 20. _____

Practice Set 9.4
Rationalizing the Denominator

Rationalize each denominator and simplify if possible.

1. $\dfrac{\sqrt{1}}{\sqrt{5}}$

1. _____

2. $\dfrac{6}{\sqrt{3}}$

2. _____

3. $\sqrt{\dfrac{3}{11}}$

3. _____

4. $\dfrac{\sqrt{a}}{\sqrt{b}}$

4. _____

5. $\sqrt{\dfrac{5}{8}}$

5. _____

6. $\dfrac{5}{\sqrt{12}}$

6. _____

7. $\sqrt{\dfrac{9}{x}}$

7. _____

8. $\dfrac{2}{\sqrt{24}}$

8. _____

9. $\sqrt{\dfrac{10}{3}}$

9. _____

10. $\sqrt{\dfrac{4x^2}{10}}$

11. $\dfrac{8}{6 - \sqrt{2}}$

12. $\dfrac{4}{3 - \sqrt{3}}$

13. $\dfrac{6}{\sqrt{3} - \sqrt{2}}$

14. $\dfrac{7}{\sqrt{5} - 2}$

15. $\dfrac{\sqrt{6} - \sqrt{3}}{\sqrt{6} + \sqrt{3}}$

16. $\dfrac{8 - \sqrt{2}}{\sqrt{5} - 2}$

17. $\dfrac{\sqrt{x}}{\sqrt{x} - 3}$

18. $\dfrac{\sqrt{x} + 4}{\sqrt{x} - 3}$

Practice Set 9.5
Radical Equations

Solve each radical equation. If the equation has no solution, so state.

1. $\sqrt{x} = 8$

    1. _____

2. $\sqrt{x} - 12 = 0$

    2. _____

3. $\sqrt{3y+4} = 7$

    3. _____

4. $\sqrt{a+100} = 25$

    4. _____

5. $\sqrt{x} + 7 = 0$

    5. _____

6. $\sqrt{3x} - 6 = 0$

    6. _____

7. $\sqrt{8x+1} = x + 2$

    7. _____

8. $\sqrt{3x-8} + 1 = 3$

    8. _____

9. $\sqrt{x+2} + 9 = 14$

    9. _____

10. $\sqrt{x-5}+5=9$ 10. _____

11. $x-5=\sqrt{4x+1}$ 11. _____

12. $4\sqrt{x}=20$ 12. _____

13. $5\sqrt{x}-6=29$ 13. _____

14. $\sqrt{x+3}=x-3$ 14. _____

15. $\sqrt{3x+4}-\sqrt{x+5}=1$ 15. _____

16. $\sqrt{x-1}+\sqrt{5x-1}=2$ 16. _____

17. $\sqrt{3x+1}+3=x$ 17. _____

18. $\sqrt{x+4}=2-\sqrt{3x}$ 18. _____

Name _____   Date _____

Practice Set 9.6
Rational Exponents

Rewrite each expression in radical form. Then simplify.

1. $25^{\frac{1}{2}}$

1. _____

2. $8^{\frac{2}{3}}$

2. _____

3. $16^{\frac{3}{2}}$

3. _____

4. $4^{\frac{-1}{2}}$

4. _____

5. $27^{\frac{-1}{3}}$

5. _____

6. $121^{\frac{1}{2}}$

6. _____

7. $\left(\frac{4}{25}\right)^{\frac{1}{2}}$

7. _____

8. $\left(\frac{1}{8}\right)^{\frac{-2}{3}}$

8. _____

9. $81^{\frac{3}{4}}$

9. _____

10. $125^{-\frac{1}{3}}$

10. _____

11. $1000^{\frac{2}{3}}$

11. _____

12. $25^{\frac{-1}{2}}$

12. _____

13. $-16^{\frac{3}{4}}$

13. _____

14. $49^{\frac{3}{2}}$

14. _____

15. $\left(\dfrac{9}{16}\right)^{\frac{-3}{2}}$

15. _____

16. $(-27)^{\frac{2}{3}}$

16. _____

17. $16^{\frac{1}{2}} + 8^{\frac{1}{3}}$

17. _____

18. $25^{\frac{3}{2}} - 27^{\frac{2}{3}}$

18. _____

19. $16^{\frac{1}{2}} \cdot 16^{\frac{1}{2}}$

19. _____

20. $49^{\frac{-1}{2}} \cdot 49^{\frac{-1}{2}}$

20. _____

Name _____   Date _____

Practice Set 10.1
Solving Quadratic Equations by the Square Root Property

Solve each quadratic equation by the square root property. If possible, simplify radicals or rationalize denominators.

1. $x^2 = 64$ 1. _____

2. $x^2 = 121$ 2. _____

3. $x^2 = 17$ 3. _____

4. $5x^2 = 45$ 4. _____

5. $16y^2 = 49$ 5. _____

6. $2x^2 - 1 = 49$ 6. _____

7. $2x^2 - 7 = 0$ 7. _____

8. $(x - 4)^2 = 81$ 8. _____

9. $(x + 1)^2 = 1$ 9. _____

10. $(x - 7)^2 = 24$ 10. _____

Solve each quadratic equation by first factoring the perfect square trinomial on the left side. Then apply the square root property. Simplify radicals, if possible.

11. $x^2 - 4x + 4 = 9$ 11. _____

12. $x^2 + 10x + 25 = 8$ 12. _____

121

13. $x^2 - 6x + 9 = 7$ 13. _____

14. $x^2 + 2x + 1 = 20$ 14. _____

Use the Pythagorean Theorem to find the missing side of the triangle with the lengths of the other two sides given. Express the answer in radical form and simplify, if possible.

15.   a = 3 meters   b = 4 meters   c = ?   15. _____

16.   a = ?   b = 12 ft.   c = 13 ft.   16. _____

17.   a = 12 in.   b = ?   c = 20 in.   17. _____

18.   One leg 6 inches; hypotenuse 10 inches. Find the other leg.   18. _____

19.   One leg 5 meters; hypotenuse 10 meters. Find the other leg.   19. _____

Find the distance between each pair of points. Express answer in simplest radical form.

20.   (2, 5)(–3, 6)   20. _____

21.   (1, 4)(–1, 8)   21. _____

22.   (0, 7)(5, –2)   22. _____

23.   (3, 4)(–1, 3)   23. _____

24.   (0, –2)(1, –4)   24. _____

25.   (–5, –2)(1, –3)   25. _____

Practice Set 10.2
Solving Quadratics by Completing the Square

Complete the square for each binomial. Then factor the resulting perfect square trinomial.

1. $x^2 + 8x$ 1. _____

2. $x^2 - 10x$ 2. _____

3. $x^2 + 16x$ 3. _____

4. $x^2 - \dfrac{3}{4}x$ 4. _____

5. $x^2 + 5x$ 5. _____

6. $x^2 - 3x$ 6. _____

Solve each equation by completing the square.

7. $x^2 + 6x + 2 = 0$ 7. _____

8. $x^2 + 6x = -8$ 8. _____

9. $x^2 + 3x = 4$ 9. _____

10. $x^2 = 2x + 2$ 10. _____

11. $x^2 - 10x + 2 = 0$ 11. _____

12. $2x^2 + 7x = 4$ 12. _____

13. $3x^2 + 2x = 4$ 13. _____

14. $3x^2 = -6x - 2$ 14. _____

15. $2x^2 - 2x = 5$ 15. _____

Name _____   Date _____

Name _____   Date _____

Practice Set 10.3
Using the Quadratic Formula

Solve each equation using the quadratic formula. Simplify irrational solutions, if possible.

1. $x^2 + 6x + 9 = 0$ 1. _____

2. $x^2 + 4x - 12 = 0$ 2. _____

3. $2x^2 + 7x = -6$ 3. _____

4. $3x^2 + 2x = 5$ 4. _____

5. $2x^2 + 9x + 4 = 0$ 5. _____

6. $x^2 - 41 = 0$ 6. _____

7. $5x^2 + 2x - 1 = 0$ 7. _____

8. $x^2 + 7x = 3$ 8. _____

9. $3x^2 + 1 = 5x$ 9. _____

10. $x^2 - 6x + 5 = 0$ 10. _____

11. $x^2 + 8x + 7 = 0$ 11. _____

12. $2x^2 = -5x + 5$ 12. _____

Name _____   Date _____

Solve each equation by the method of your choice. Simplify irrational solutions, if possible.

13. $4x^2 = 100$ 13. _____

14. $(3x + 2)^2 = 12$ 14. _____

15. $3x^2 = 120$ 15. _____

16. $(5x - 1)^2 = 32$ 16. _____

17. $2x^2 - 4x + 1 = 0$ 17. _____

18. $x^2 + 4x = 2$ 18. _____

19. $5x^2 - x - 2 = 0$ 19. _____

20. $\dfrac{1}{2}x^2 - \dfrac{1}{3}x = \dfrac{5}{6}$ 20. _____

Practice Set 10.4
Imaginary Numbers as Solutions of Quadratic Equations

Express each number in terms of i.

1. $\sqrt{-25}$ 1. _____

2. $\sqrt{-121}$ 2. _____

3. $-\sqrt{-24}$ 3. _____

4. $\sqrt{-18}$ 4. _____

5. $\sqrt{-1}$ 5. _____

6. $\sqrt{0}$ 6. _____

7. $\sqrt{-44}$ 7. _____

8. $\sqrt{-30}$ 8. _____

9. $4 + \sqrt{-7}$ 9. _____

10. $7 + \sqrt{-9}$ 10. _____

Solve each quadratic equation using the square root property. Express imaginary solutions in $a + bi$ form.

11. $(y-2)^2 = -25$ 11. _____

12. $(x+3)^2 = -8$ 12. _____

13. $(x-1)^2 = -60$ 13. _____

Solve each quadratic equation using the quadratic formula.

14. $x^2 + 10x + 28 = 0$ 14. _____

15. $x^2 - 6x = -14$ 15. _____

16. $16x^2 - 8x + 7 = 0$ 16. _____

17. $3x^2 - 5x = -7$ 17. _____

18. $x^2 = x - 4$ 18. _____

19. $4x^2 - 3x + 5 = 0$ 19. _____

20. $5x^2 = x - 1$ 20. _____

Practice Set 10.5
Graphs of Quadratic Equations

Determine if the parabola, whose equation is given, opens upward or downward.

1. $y = x^2 - 3x + 1$ 1. _____

2. $y = -2x^2 + 5x - 2$ 2. _____

Find the x-intercepts for the parabola whose equation is given.

3. $y = x^2 - 5x + 4$ 3. _____

4. $y = -x^2 - 7x - 12$ 4. _____

Find the y-intercept for the parabola whose equation is given.

5. $y = x^2 - 7x + 5$ 5. _____

6. $y = -x^2 - 2x + 4$ 6. _____

Find the vertex for the parabola whose equation is given.

7. $y = x^2 + 2x - 3$ 7. _____

8. $y = -x^2 - 4x - 3$ 8. _____

Graph the parabola whose equation is given.

9. $y = -x^2 - 2x + 3$ 10. $y = x^2 - 6x + 5$

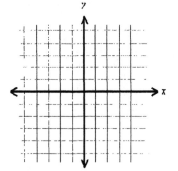

11. $y = 2x^2 + 6x + 4$

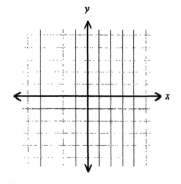

12. $y = x^2 - 1$

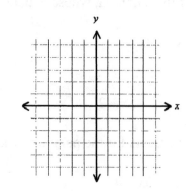

13. $y = -x^2 + 8$

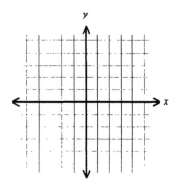

14. $y = 3x^2 + 4x + 1$

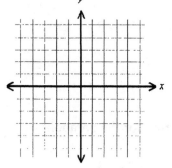

15. $y = -x^2 + 4$

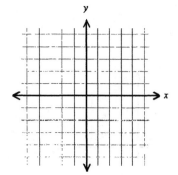

16. $y = -x^2 - 4x + 5$

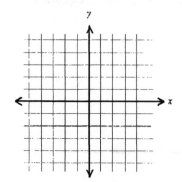

Name _____   Date _____

Practice Set 10.6
Introduction to Functions

In exercises 1-4, (a) determine whether each relation is a function, (b) give the domain and (c) range for each relation.

1.   {(1, 2)(4, 5)(6, 9)}   1a. _____

                                                                 b. _____

                                                                 c. _____

2.   {(3,1)(3,2)(3, 3)(3, 4)}   2a. _____

                                                                 b. _____

                                                                 c. _____

3.   {(5, 1)(5, 7)(5, 11)}   3a. _____

                                                                 b. _____

                                                                 c. _____

4.   {(1, 2)(2, 3)(3, 4)(4, 5)}   4a. _____

                                                                 b. _____

                                                                 c. _____

Evaluate each function at the given values.

5. $f(x) = 7 - x$

        a. $f(-1)$ 5a. _____

        b. $f(0)$ b. _____

        c. $f(3)$ c. _____

6. $g(x) = x^2 + 4$

    a. $f(-2)$ 6a. _____

    b. $f(0)$ b. _____

    c. $f(4)$ c. _____

7. $f(x) = |-2x + 5|$

    a. $f(-4)$ 7a. _____

    b. $f(0)$ b. _____

    c. $f(6)$ c. _____

8. $f(x) = 4x + 3$

    a. $f(-1)$ 8a. _____

    b. $f(0)$ b. _____

    c. $f(5)$ c. _____

Use the vertical line test to identify graphs in which y is a function of x.

9.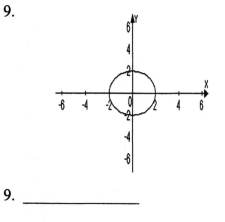

10.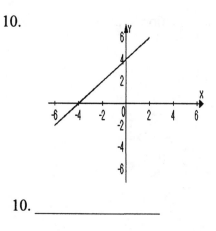

9. _____   10. _____

11.

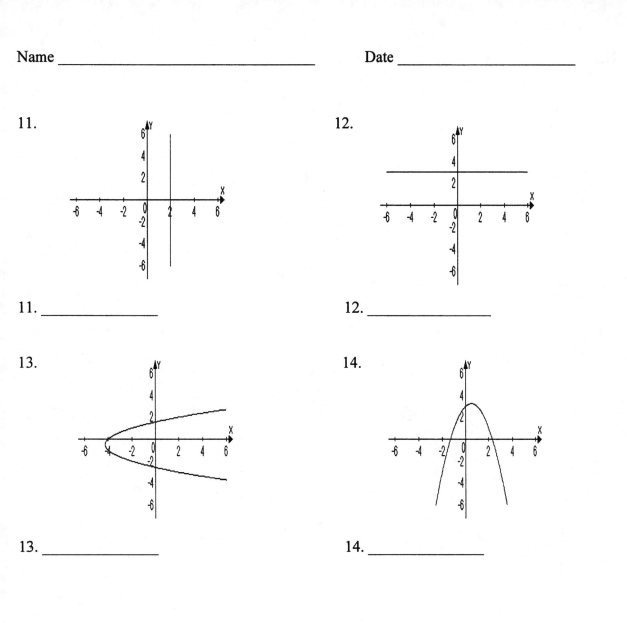

11. _____

12. _____

13. _____

14. _____

Name _____   Date _____

Practice Set Answers Chapter 1

1.1

1. 9 3. 2 5. $x - 5$ 7. $x - 8$ 9. $2x - 4 = 6$ 11. $6x + 3$ 13. $8 \div x + 9$ or $\dfrac{8}{x} + 9$

15. yes 17. yes 19. yes

1.2

1. $\dfrac{37}{8}$ 3. $8\dfrac{4}{5}$ 5. $2 \cdot 2 \cdot 2 \cdot 2 \cdot 5$ 7. $\dfrac{3}{2}$ 9. $\dfrac{1}{3}$ 11. $\dfrac{3}{7}$ 13. 21 15. $\dfrac{93}{50}$ or $1\dfrac{43}{50}$ 17. $\dfrac{3}{10}$

19. $9\dfrac{13}{24}$

1.3

1. 0.625 3. 0.45 5. $-18, 0, 32$ 7. $15, 1$ 9. true 11. false 13. < 15. < 17. < 19. <

1.4

1a. 2 terms b. none 3a. 4 terms b. $5y$ and $2y$ 5. Associative property of multiplication
7. Distributive property 9. Distributive property 11. $8y$ 13. $8a + 40$
15. $6x + 5$ 17a. $8x - 2x$ b. $6x$ 19a. $2x + 16 - 5$ b. $2x + 11$

1.5

1. 4 3. -24 5. -2.2 7. -58 9. $-\dfrac{1}{2}$ 11. 5 13. 16 15. 3 17. a 19. 0

1.6

1. -6 3. -17.3 5. 70 7. $\dfrac{-8}{9}$ 9. -60 11. 20 13. 235 15. -212 17. $9m - 22$ 19. $x - 4$

1.7

1. -12 3. 0 5. -60 7. $\dfrac{9}{10}$ 9. $2a$ 11. -0.096 13. 0 15. -11.1 17. 24 19. $-5y + 4$

1.8

1. -49 3. 64 5. 125 7. cannot be simplified 9. $-2x + 52$ 11. 17 13. 0 15. 7 17. 27
19. -47

Practice Set Answers Chapter 2

2.1

1. linear   3. not linear   5. not linear   7. $\{-15\}$ 9. $\{6\}$ 11. $\{0\}$ 13. $\left\{\dfrac{1}{2}\right\}$ 15. $\{3\}$ 17. $\{5\}$
19. $\{-4\}$

2.2

1. $\{-12\}$ 3. $\{18\}$ 5. $\{5\}$ 7. $\left\{-\dfrac{1}{2}\right\}$ 9. $\{-35\}$ 11. $\{6\}$ 13. $\{10\}$ 15. $\{7\}$ 17. $\{-2\}$ 19. $\left\{\dfrac{11}{2}\right\}$

2.3

1. $\{-3\}$ 3. $\{11\}$ 5. $\{0\}$ 7. $\{-4\}$ 9. $\{4\}$ 11. $\{-7\}$ 13. $\{1\}$ 15. $\{3\}$ 17. $\{-6\}$ 19. $\varnothing$

2.4

1. $a = P - b - c$ 3. $x = \dfrac{y-b}{m}$ 5. $A = \dfrac{bh}{2}$ 7. $R = \dfrac{PV}{nT}$ 9. 0.15   11. 0.0075   13. 1.5%
15. 7   17. 250   19. 15%

2.5

1a. $x - 5 = 11$ b. $x = 16$ 3a. $2x + 3 = 5$ b. $x = 1$ 5a. $x + 2 = 3x$ b. $x = 1$ 7a. $11 = x - 4$ b. $x = 15$
9a. $6x - 2 = 16$ b. $x = 3$ 11a. $8x + 9 = 1$ b. $x = -1$ 13a. $2x + 6 = x - 11$ b. $x = -17$
15a. $4x - 8 = 2x + 4$ b. $x = 6$

2.6

1a.   b. $[-4, \infty)$

3a.   b. $(3, \infty)$

5a.   b. $(-2, 4]$

7a.   b. $(3, \infty)$

9a.   b. $(-5, \infty)$

11a.   b. $[1, \infty)$

13a.   b. $(-\infty, 4]$

15a.   b. $(3, \infty)$

17a.   b. $(-\infty, -1)$

19a.   b. $(-3, \infty)$ 21. $(-\infty, \infty)$

Practice Set Answers Chapter 3

3.1

1. $150 3. $4,500 @ 4%, $5,500 @ 5% 5. $900 @ 8%, $1,200 @ 9.5% 7. 71.25 ml

9. 8 gallons of 30%, 8 gallons of 70% 11. 5.5 hours 13. 6 hours 15. 30 mph and 40 mph

3.2

1. $\dfrac{2}{3}$ 3. $\dfrac{19}{4}$ 5. Yes 7. $a = 2.625$ 9. $p = 13.75$ 11. $x = 6$ 13. $x = 4$ 15. $62.50

17. 90 males, 315 females

3.3

1. 112 in.^{2} 3. 25 in.^{2} 5. 9π in.^{2} ; 28 in.^{2} 7. 10π cm; 31 cm 9. 9 feet 11. 15.625 in.^{3}

13. 6.75π ft.^{3} ; 21 ft.^{3} 15. 29°, 58°, 93° 17. 17° 19. 44° and 136°

Practice Set Answers Chapter 4

4.1

1 – 3

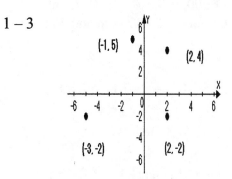

1. quadrant I   3. quadrant II
5. (3, 1)   7. (–1. –1)
9a. No b. No c. Yes

11. (–2, –2)(–1, –1)(0, 0)(1, 1)(2, 2)

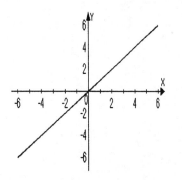

13. (–2, -7)(–1, –4)(0, –1)(1, 2)(2, 5)

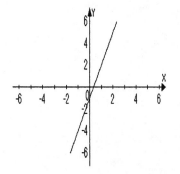

15. (–4, 3)(–2, 2)(0, 1)(2, 0)(4, –1)

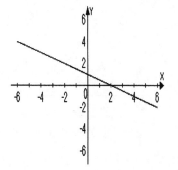

1. $(3, 0)(0, -5)$ 3. $(0, 0)(0, 0)$ 5. $\left(\dfrac{7}{4}, 0\right)\left(0, \dfrac{-7}{3}\right)$

7.

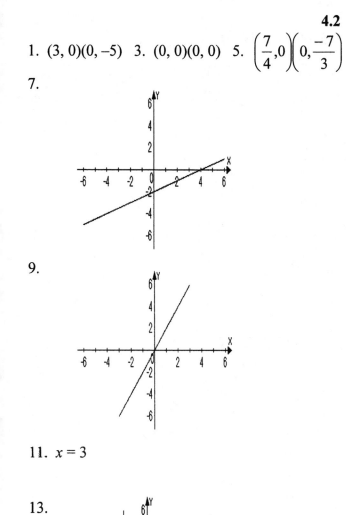

9.

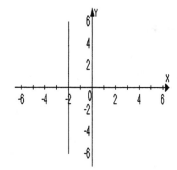

11. $x = 3$

13.

4.3

1. $m = \dfrac{3}{1}$, rises   3. $m = \dfrac{-13}{5}$, falls   5. $m = -2$, falls   7. m is undefined, vertical

9. $m = -\dfrac{2}{3}$ 11. m is undefined   13. perpendicular   15. neither.

1a. $m = 2$ b. y-intercept $(0, -4)$ 3a. $m = 0$ b. y-intercept $(0, 3)$

5a. $m = \dfrac{2}{5}$ b. y-intercept $(0, 3)$ c.

7a. $m = \dfrac{-3}{4}$ b. y-intercept $(0, 0)$ c.

9a. $m = \dfrac{1}{4}$ b. y-intercept $(0, 2)$ c.

11a. $m = \dfrac{-5}{3}$ b. y-intercept $(0, -4)$ c.

13. negative 15. undefined

4.5

1a. $y - 1 = -2(x - 3)$ b. $y = -2x + 7$ 3a. $y - 1 = \dfrac{-2}{3}(x - 0)$ b. $y = \dfrac{-2}{3}x + 1$

5a. $y - 0 = -4(x - 0)$ b. $y = -4x$ 7a. $y - 4 = 1(x - 0)$ or $y - 7 = 1(x - 3)$ b. $y = x + 4$

9a. $y - 4 = \dfrac{1}{2}(x + 2)$ b. $y = \dfrac{1}{2}x + 5$

4.6

1a. Yes   b. Yes   c. No   3a. Yes   b. No   c. Yes

5.

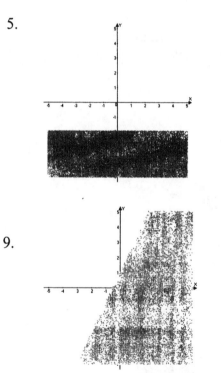

7.

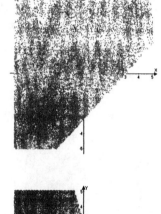

9.

11.

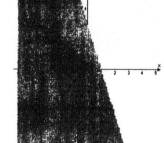

5.1

1. Yes   3. No   5. $\{(-1, -3)\}$ 7. $\{(4, 1)\}$ 9. No Solution   11. $\{(5, -5)\}$ 13. $\{(1, -4)\}$

5.2

1. $\{(-2, 1)\}$ 3. $\left\{\left(\dfrac{7}{19}, \dfrac{-4}{19}\right)\right\}$ 5. $\{(4, -2)\}$ 7. No Solution   9. $\{(2, 1)\}$

11. infinitely many solutions $\{(x, y) \mid 4x - y = 10\}$ or $\{(x, y) \mid 8x - 2y = 20$ 13. $\{(5, -8)\}$
15. $\{(4, 2)\}$ 17. $\{(5, 1)\}$

5.3

1. $\{(-4, -6)\}$ 3. $\{(-5, 2)\}$ 5. $\{(-4, 3)\}$ 7. $\{(0,1)\}$ 9. $\{(4, 1)\}$ 11. $\{(-5, -3)\}$
13. No Solution   15. $\{(1, 9)\}$ 17. $\{(5, 3)\}$

5.4

1. 4 and 7   3. -4 and -12 5. width = 15 centimeters, length = 24 centimeters
7. width = 100 feet, length = 150 feet   9. 45 of the 30¢ stamps, 55 of the 39¢ stamps
11. 9 hrs. at coffee shop, 15 hrs. cleaning offices

5.5

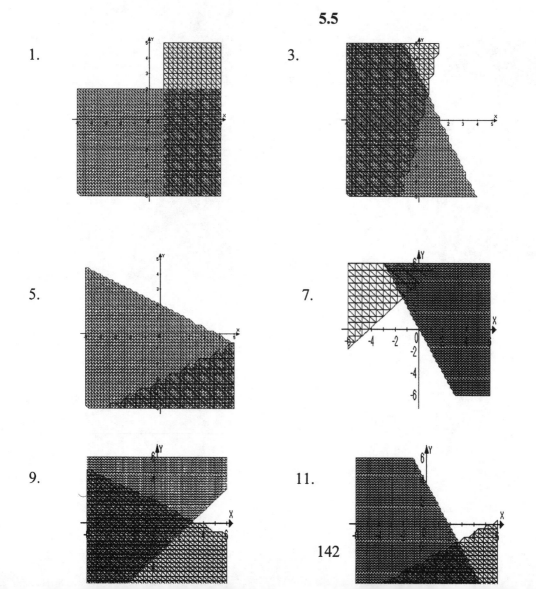

1.

3.

5.

7.

9.

11.

Practice Set Answers Chapter 6

6.1

1. binomial, 1   3. monomial, 3   5. trinomial, 2   7. $-2x+10$ 9. $-12x^2-2x+9$
11. $3x^3+2x^2-10$ 13. $11x-1$ 15. $-x^2-9x+2$ 17. $9x^3+2x^2-11x+3$

6.2

1. x^5 3. a^7 5. 3^{12} 7. $(-10)^{16}$ 9. $27x^6$ 11. $16y^{12}$ 13. $30x^5$ 15. $16x^5$
17. $12x^4-16x^3+20x^2$ 19. $12x^2-5x-2$ 21. $6x^2+19x-7$ 23. $4x^2-27x-7$
25. x^4-9x^2+6x-1

6.3

1. $x^2+10x+21$ 3. $15x^2+2x-8$ 5. $28x^4-6x^2-10$ 7. $81x^2-1$ 9. $16x^2-\dfrac{1}{16}$

11. $4x^2+20x+25$ 13. x^2-2x+1 15. $9y^2-2y+\dfrac{1}{9}$ 17. x^2+4x+4 19. $9y^2-4$

6.4

1. 31   3a. 4, –3, 4   b. 5, 3, 6   c. 6   5a. 13, 3, –5, 1   b. 5, 7, 7, 10   c. 10   7. $-x^2y-4xy-5$
9. $6a^3+8a^2b-7ab^3+6b^4$ 11. $-a^2b^4+8ab^2-2ab$ 13. $30x^7y^6$ 15. $9x^2-24xy+16y^2$
17. $12x^5y-8x^4y^2+20x^6y^5$ 19. $4x^2-25y^2$

6.5

1. 5^5 3. x^3y^9 5. 1   7. –1   9. $\dfrac{x^2}{16}$ 11. $\dfrac{36x^8}{49}$ 13. $5x^7$ 15. $6ab^3c^7$ 17. $8x^5+3x^2$
19. $-4x^3y^2+6xy^3-2y^4$

6.6

1. $x-2$ 3. $x-4+\dfrac{8}{x+2}$ 5. $2x^2-x-2$ 7. $2x^2+8x-7-\dfrac{19}{3x-2}$ 9. x^3+3x^2+2x-2

6.7

1. $\dfrac{1}{3^3}=\dfrac{1}{27}$ 3. $\dfrac{1}{3}+\dfrac{1}{4}=\dfrac{7}{12}$ 5. x^7 7. x^{11} 9. 31,000   11. 0.0532   13. 1.58×10^4
15. 5.7×10^{-3} 17. 4×10^7 19. 4×10^9

Practice Set Answers Chapter 7

7.1

1. $9(x+1)$ 3. $x^2(x+4)$ 5. $3(6x^4+3x^2+1)$ 7. cannot be factored 9. $3x^2y^2(2y+5x^2)$
11. $(x-4)(x+3)$ 13. $(x+9)(7x-1)$ 15. $(7+a)(2-b)$ 17. $(m+8)(n+1)$
19. $(a+3)(2+b)$

7.2

1. $(x+2)(x+4)$ 3. $(x+4)(x-1)$ 5. $(y-6)(y-2)$ 7. $(a-7)(a+2)$ 9. $(m-6)(m-4)$
11. $5(x+4)(x+1)$ 13. $(x-7)(x-1)$ 15. $(x+5y)(x-2y)$ 17. $2y(y+2)(y+6)$
19. $(x-7)(x-8)$

7.3

1. $(3y+1)(3x+4)$ 3. $(2x-5)(2x-3)$ 5. $(2a-3)(2a-3)$ 7. $(2x+3)(x+5)$
9. $(2x+7)(x-1)$ 11. $(5x+2)(2x+3)$ 13. $(6a-5)(3a+2)$ 15. $(4x-5)(4x+3)$
17. $(10a-3)(4a+7)$ 19. $3a(a-8)(a+2)$

7.4

1. $(x+4)(x-4)$ 3. prime 5. $(x+8)^2$ 7. $3(2x+y)(2x-y)$ 9. $(2x-3y)(2x-3y)$
11. $(5a+1)^2$ 13. $(x+1)(x-1)$ 15. $(y-7)^2$ 17. $(3m-8)^2$ 19. $(x^3+5y)(x^3-5y)$

7.5

1. $(x+14)(x-14)$ 3. $(x-7)(y+2)$ 5. $2(3m+5)(3m-5)$ 7. $(x+3a)(x-2b)$
9. $(x+4y)(x^2-4xy+16y^2)$ 11. $(x+5)(x+4)(x-4)$ 13. $2(x+4)(x-4)$
15. $(4x+5y)(5x+y)$ 17. $(2a+5)(4a^2-10a+25)$ 19. $(3x-7)(2x+5)$

7.6

1. $\{0,-4\}$ 3. $\{3,-7\}$ 5. $\{-4,-3\}$ 7. $\{5,-4\}$ 9. $\{-3,6\}$ 11. $\{0,4\}$ 13. $\{-5,2\}$
15. $\left\{\dfrac{1}{4},\dfrac{-1}{2}\right\}$ 17. $\{-5,1\}$ 19. width = 4 inches, length = 8 inches

Practice Set Answers Chapter 8

8.1

1. $x = 0$ 3. defined for all real numbers   5. $2(x+2)$ or $2x+4$ 7. 3   9. $\dfrac{t+2}{2(t-4)}$ 11. $\dfrac{-5}{y+6}$

13. cannot be simplified   15. $\dfrac{1}{x-1}$ 17. x^2+3x+9 19. $\dfrac{4}{x+1}$

8.2

1. $\dfrac{3x-12}{5x+10}$ 3. $\dfrac{1}{(a+1)(a-3)}$ 5. $\dfrac{x-3}{2}$ 7. $\dfrac{x^2+3x+9}{x-3}$ 9. $\dfrac{5}{y+3}$ 11. $\dfrac{5x}{36}$ 13. $\dfrac{2x}{3}$

15. $\dfrac{y-2}{y+2}$ 17. $\dfrac{2(a+5)}{(a-2)}$ 19. $\dfrac{1}{3}$

8.3

1. $\dfrac{6x}{7}$ 3. $\dfrac{-2x+5}{4}$ 5. $x-5$ 7. $\dfrac{3x^2+4x-15}{2x+1}$ 9. $\dfrac{-3x+18}{x+7}$ 11. $\dfrac{3x+3}{x-3}$ 13. $\dfrac{3}{x+y}$

15. $\dfrac{1}{x-5}$

8.4

1. $40x^3$ 3. $5x(x+2)(x-2)$ 5. $\dfrac{4}{3}$ 7. $\dfrac{3x-4}{4}$ 9. $\dfrac{5}{y-4}$ 11. $\dfrac{y^2-3y+5}{(y+3)(y-3)}$

13. $\dfrac{x+38}{(x+5)(x-4)(x+3)}$ 15. $\dfrac{y+3}{y+1}$

8.5

1. $\dfrac{9}{4}$ 3. $\dfrac{5x+16}{2x+24}$ 5. $\dfrac{xy+6}{x}$ 7. $\dfrac{x+3}{2x+x^2}$ 9. $-\dfrac{8}{5}$ 11. $\dfrac{b+2a}{3+a^2}$ 13. $\dfrac{x^2y-3x}{y-8x}$

15. $\dfrac{(x-2)(x+1)}{(x-1)(3x+4)}$

8.6

1. {72}   3. {9}   5. {6}   7. {20}   9. Ø   11. {−6, −1}   13. {4}

8.7

1. 5 mph   3. 3 mph   5. $2\dfrac{8}{11}$ hrs.   7. 15 ft.   9. $y = 60$ 11. $y = 2$

Practice Set Answers Chapter 9

9.1

1. 8   3. –5   5. –10   7. $\dfrac{7}{10}$ 9. –1   11. 4   13. Not a real number   15. –10   17. 8.944

19. Not a real number

9.2

1. $3\sqrt{2}$ 3. $2x\sqrt{5}$ 5. $6a\sqrt[4]{2}$ 7. $3y\sqrt[3]{2}$ 9. $2y\sqrt[6]{7}$ 11. $10x^7\sqrt{3x}$ 13. $6a^2\sqrt{3a}$

15. $2x^3\sqrt{5x}$ 17. $3x$ 19. $2y\sqrt[4]{2}$

9.3

1. $9\sqrt{6}$ 3. $6\sqrt{5}$ 5. $7\sqrt{3}$ 7. $15\sqrt{3y}$ 9. $3\sqrt{2}$ 11. $\sqrt{30}+2\sqrt{6}$ 13. $2\sqrt{5}+\sqrt{30}$

15. $x+2\sqrt{2x}+2$ 17. 48   19. $24\sqrt{2}-20\sqrt{6}+4\sqrt{3}-10$

9.4

1. $\dfrac{\sqrt{5}}{5}$ 3. $\dfrac{\sqrt{33}}{11}$ 5. $\dfrac{\sqrt{10}}{4}$ 7. $\dfrac{3\sqrt{x}}{x}$ 9. $\dfrac{\sqrt{30}}{3}$ 11. $\dfrac{24+4\sqrt{2}}{17}$ 13. $6\sqrt{3}+6\sqrt{2}$

15. $3-2\sqrt{2}$ 17. $\dfrac{x+3\sqrt{x}}{x-9}$

9.5

1. {64}   3. {15}   5. No Solution   7. {3, 1}   9. {23}   11. {12}   13. {49}   15. {4}

17. {8}

9.6

1. $\sqrt{25}$; 5   3. $\left(\sqrt{16}\right)^3$; 64   5. $\dfrac{1}{\sqrt[3]{27}}$; $\dfrac{1}{3}$ 7. $\sqrt{\dfrac{4}{25}}$; $\dfrac{2}{5}$ 9. $\left(\sqrt[4]{81}\right)^3$; 27   11. $\left(\sqrt[3]{1000}\right)^2$; 100

13. $-\left(\sqrt[4]{16}\right)^3$; –8   15. $\left(\sqrt{\dfrac{16}{9}}\right)^3$; $\dfrac{64}{27}$ 17. $\sqrt{16}+\sqrt[3]{8}$; 6   19. $\sqrt{16}\cdot\sqrt{16}$; 16

Practice Set Answers Chapter 10

10.1

1. ± 8 3. $\pm\sqrt{17}$ 5. $\pm\dfrac{7}{4}$ 7. $\dfrac{\pm\sqrt{14}}{2}$ 9. $0, -2$ 11. $5, -1$ 13. $3\pm\sqrt{7}$ 15. $c = 5$ meters

17. 16 in.   19. $5\sqrt{3}$ meters   21. $2\sqrt{5}$ 23. $\sqrt{17}$ 25. $\sqrt{37}$

10.2

1. $-2x^2 + 8x + 16 = (x+4)^2$ 3. $x^2 + 16x + 64 = (x+8)^2$ 5. $x^2 + 5x + \dfrac{25}{4} = \left(x+\dfrac{5}{2}\right)^2$

7. $\left\{-3\pm\sqrt{7}\right\}$ 9. $\{1, -4\}$ 11. $\left\{5\pm\sqrt{23}\right\}$ 13. $\left\{\dfrac{-1\pm\sqrt{13}}{3}\right\}$ 15. $\left\{\dfrac{1\pm\sqrt{11}}{2}\right\}$

10.3

1. $\{-3\}$ 3. $\left\{-\dfrac{3}{2}, -2\right\}$ 5. $\left\{-4, -\dfrac{1}{2}\right\}$ 7. $\left\{\dfrac{-1\pm\sqrt{6}}{5}\right\}$ 9. $\left\{\dfrac{5\pm\sqrt{13}}{6}\right\}$ 11. $\{-7, -1\}$

13. $\{-5, 5\}$ 15. $\left\{-2\sqrt{10}, 2\sqrt{10}\right\}$ 17. $\left\{\dfrac{2\pm\sqrt{2}}{2}\right\}$ or $\left\{1\pm\dfrac{\sqrt{2}}{2}\right\}$ 19. $\left\{\dfrac{1\pm\sqrt{41}}{10}\right\}$

10.4

1. $5i$ 3. $-2i\sqrt{6}$ 5. i 7. $2i\sqrt{11}$ 9. $4+i\sqrt{7}$ 11. $\{2\pm 5i\}$ 13. $\left\{1\pm 2i\sqrt{15}\right\}$

15. $\left\{3\pm i\sqrt{5}\right\}$ 17. $\dfrac{5\pm i\sqrt{59}}{6}$ 19. $\dfrac{3\pm i\sqrt{71}}{8}$

10.5

1. up   3. 4 and 1   5. $(0, 5)$ 7. $(-1, -4)$

9.

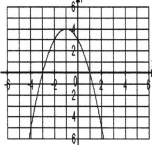

11.

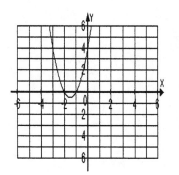

13.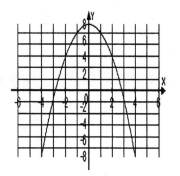

15.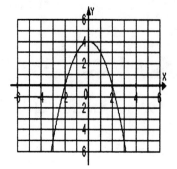

10.6

1a. function b. domain: {1, 4, 6} c. range: {2, 5, 9}

3a. not a function b. domain {5} c. range: {1, 7, 11} 5a 8 b. 7 c. 4 7a. 13 b. 5 c. 7

9. not a function 11. not a function 13. not a function

148